杰罗梅·布鲁奈特与艾瑞克·索尼尔

Urban Sites

都市场所

Urban Sites

——杰罗梅·布鲁奈特与艾瑞克·索尼尔建筑作品集

Jérôme Brunet Eric Saunier

[德] 塞巴斯蒂安·雷德克 著
王 葳 译

中国建筑工业出版社

著作权合同登记图字：01-2003-4468号

图书在版编目（CIP）数据

都市场所　杰罗梅·布鲁奈特与艾瑞克·索尼尔建筑作品集/
(德) 塞巴斯蒂安·雷德克著；王葳译.—北京：中国建筑工业出版社，2008

ISBN 978-7-112-10260-0
I.都… II.①塞…②王… III.建筑设计-作品集-法国-现代
IV.TU206

中国版本图书馆CIP数据核字（2008）第118150号

Jérôme Brunet / Eric Saunier / Urban Site / Sebastian Redecke

责任编辑：戚琳琳　段　宁
责任设计：董建平
责任校对：梁珊珊　陈晶晶

都市场所
——杰罗梅·布鲁奈特与艾瑞克·索尼尔建筑作品集
[德] 塞巴斯蒂安·雷德克　著
王葳　译

中国建筑工业出版社 出版、发行（北京西郊百万庄）
各地新华书店、建筑书店经销
北京美光制版有限公司制版
北京盛通印刷股份有限公司印刷

*

开本：880×1230毫米　1/16　印张：9　字数：288千字
2009年5月第一版　　2009年5月第一次印刷
定价：78.00元
ISBN 978-7-112-10260-0
(17063)

(邮政编码　100037)

Contents

目录

碰撞

Encounters

尼姆（Nîmes）拥有一个辉煌的古罗马圆形剧场。它始建于公元1世纪的弗兰维尔时代，规模宏大，长103m，宽101m。有韵律感的巨大拱廊构成了剧场典型的外部形象，位于下部的拱廊被饰以壁柱，而上部则是由多立克柱支撑。这是至今保存最完好的罗马剧场，可以容纳7000个座位， Labfac建筑师为它重新设计了一个玻璃屋顶，剧场借此重新焕发出昔日的光彩，用于夜间演出和网球比赛。从北到西，剧场建筑群成为古镇不可或缺的一部分。许多窄巷都汇入环绕剧场北侧的林荫广场。在这个高密度的城市中只有维克多·雨果林荫大道才给人带来宽阔的空间感觉，它一直向南延伸到另外一个古老的建筑——卡利神殿(Maison Carrée)，一个坐落在基座上保存完好的神殿。再往下是卡利艺术中心——

Nîmes has its magnificent Roman amphitheatre. The oval structure, erected in the first century AD in Flavian times, is a notable 133 metres long and 101 metres wide. Its facades are characterised by the typical rhythm of massive rows of arcades, the lower row ordered by pilasters, the upper by engaged Doric columns. The best-preserved Roman theatre, with its 7000 seats and a new, spectacular roof by Labfac architects, is used today for large-scale events ranging from a night of opera to a tennis tournament. To the north and west, the building becomes a part of the old town with its tightly packed houses. Narrow streets unexpectedly open onto the Boulevard des Arènes that circles the big round in the north. The only exception in the dense urban tissue is the wide Boulevard Victor Hugo. It leads north to another building of antiquity, the Maison Carrée, a well-preserved podium temple, and to the Carré d'Art, the city's cultural centre designed by Norman Foster. The

◄ 尼姆 La Placette 小学的外立面

Facade detail of the elementary school " de la Placette " in Nîmes

诺曼·福斯特(Norman Foster)设计的城市文化中心。从古镇高密度的空间到圆形剧场前开阔的大广场，空间的张弛令人眩目，即使从各个不同的方位上看亦是如此。

距剧院西侧250m的地方又呈现出完全不同的景象，进入位于圣约瑟诊所后面的医院，有一段楼梯通向小礼拜堂，它前面方形小广场是古镇中最美的景致之一，它的北侧边界由一长段现代高科技感很强的铝合金玻璃幕墙外立面所界定。在这里我们与建筑师杰罗梅·布鲁奈特和艾瑞克·索尼尔的设计思想和作品初遇。

这个立面从几个层面上表达了他们典型的建筑语言，阐述了布鲁奈特和索尼尔解决设计问题的方法。首先，建筑师毫不犹豫地发展了新生的和创新的立面结构。通过进一步的观察，我们会发现，他们所设计的建筑和周围的历史文脉息息相

change from the density of the old town to the expanse of the large square with the amphitheatre is striking. Seen from different directions it is surprising again and again.

Only 250 metres to the west of the theatre there is a totally different situation. Upon entering Rue de l'Hôtel-Dieu behind Clinique Saint-Joseph, one steps onto the small Place de l'Oratoire, a triangular square in one of the most beautiful parts of the old town. On its northern edge, the square is bordered by a long, very technical looking facade of aluminium slats. This is the first, to begin with puzzling encounter with the architecture of Jérôme Brunet and Eric Saunier.

This facade speaks their architectural language on several levels. In an exemplary way, it demonstrates how Brunet and Saunier tackle a design problem. First of all, it is characteristic that the architects do not hesitate to develop a fundamentally new and innovative facade structure. Upon closer inspection, however, it also becomes clear that they nevertheless succeed in integrating the building into its surrounding ensemble, in this

通，与已有的教堂和神殿有机地融合在一起。布鲁奈特和索尼尔将对新建筑的设计和小礼拜堂的重新设计合二为一，以期创造出和谐均衡的建筑艺术作品。胸怀坚定的设计理念，他们设计了一个4层的、能容纳500个学生的学校建筑，它成为周围环境中的又一个亮点。建筑于1991年建成，由20世纪早期建成的已有部分和一个新建的部分组成。建筑师仅仅是沿南侧将原来的建筑体扩大一倍，新旧建筑的对话在中央走廊中展开，隔墙分立在走廊两侧，走廊两端各有一门分别通向两所建筑。

他们执著的设计理念在朝向广场的建筑立面中得到了淋漓尽致的展现。建筑立面由下至檐口都是由一种连续的铝合金百叶幕墙材质构成，幕墙表面呈现出时时变幻的光影。建筑师将此看作是："放大了的软百叶窗的模板"。除了女儿墙外

case the historic fabric and the adjoining church, Temple de l'Oratoire. Brunet and Saunier took the brief for the new building and the redesign of the Place de l'Oratoire to be one task demanding a harmonious balancing-act. With great resoluteness in design they created a four-storey school building for 500 pupils, establishing a new visual focus in the neighbourhood. Completed in 1991, the Ecole de la Placette comprises a building from the early 1900s and a new building. The architects simply "doubled" the old building volume on its south side. Old and new are engaged in an intense dialogue best experienced in the central corridors. Here the interior walls, with doors to the classrooms of both building parts, stand face to face.

Their resoluteness in design, however, is most apparent in the new facade facing the square. Up to the parapet of the open roofstorey, the facade is one continuous, long front made up of simple aluminium slats grouped into panels, a delicate structure producing an ever-changing play of shadows. The architects think of it as a "transcription of a greatly enlarged Venetian blind". Each framed panel with its fixed slats–except those at the level of the parapets or in

挂的幕墙，其他每一个单元的窗户都可沿中轴旋转打开。这样使广场的美景尽收眼底。

在建筑师的设计理念中最基本的构成是对于建筑文脉的考虑。由于不愿受建筑形式和建筑语言等的限制，两位建筑师的建筑理念显得大胆而冒险，他们没有着力于寻找简单或者直接的设计方法，而是将一个场所重新赋予建筑艺术的生命力。抱着这样一种大胆的创作精神，自1981年杰罗梅·布鲁奈特和艾瑞克·索尼尔从业以来，他们令人惊讶地在参加的建筑竞赛中均有所斩获。在这些竞赛中，他们不时地奉献出非同寻常、立意新颖的建筑造型，并努力将他们的设计理念变得生动和更具说服力。

在尼姆的建筑设计竞赛中，他们将新建建筑的南立面用铝合金百叶幕墙包起来，以抵消建筑在夏季中的过热反应。铝合金百叶装在玻璃幕墙的外部，这像是

front of a closed wall surface - can be pivoted around a central axis. The unobstructed view of the square is therefore available to all.

Fundamental considerations of context constitute a further theme in the architects' work. Unwilling to be categorised by building type or vocabulary, this is taking a risky position. They are not interested in searching for the easy, perhaps obvious path. They want to transform a place architecturally, want to redefine it. With their daring attitude, it is astonishing that ever since they founded their practice in 1981, Jérôme Brunet and Eric Saunier were awarded almost all commissions as the result of architectural competitions. In these competitions, time and again they presented unconventional, newly developed facades. With great dedication, they managed to present their concepts in a vivid and convincing way.

In the competition for Nîmes, they chose to clad the new building' s southern front with an aluminium structure to counteract the extreme overheating of the building in summer. The

建筑师的自述。因为在立面的双层皮中运用了最简单的几种材料，所以施工造价不会超出预算。此外，将外部的百叶窗关上，为学生营造相对隔绝的学习环境——不被外界吸引，也不会吸引外界的注意。

整个建筑立面都覆以铝合金百叶，在视觉上产生极大的冲击力，从而使其他构成元素变得黯淡无光。建筑前方朝向广场的下沉场地是专为一年级的学生设置的。这是室内空间的延伸，既为过路人带来开敞的绿色空间，也使低年级的学生能够在远离街道的、安全的室外环境中玩耍。

aluminium slat wall was set in front of the actual, mostly glazed facade in a way that seems self-explanatory to the architects. Because only simple building materials were used in this double facade, construction costs did not exceed the budget. In addition, the building's certain hermetic quality that results when the slat wall is closed prevents pupils from being too distracted by gazing out or being seen.

It would surely have been too provocative to extend the uncompromising aluminium skin across the entire length of the building. It was therefore ingeniously "played down" by other elements. A sunken playground for the first graders is open to the square in front of the building. By placing it there, an additional floor was gained, passers-by have a view of this partially planted open area, and the younger children can play in a space that is outdoors, but separated from the street.

另外一种减弱铝合金百叶在立面中所占据重要地位的方法就是将它看作是立面组成的一部分。由于底层外部缺乏室外活动空间，在第二层和第三层平面上设计了两处大型开放的屋顶花园。索膜结构纵身穿过整个新建建筑，飞跨于整个新建筑上，遮蔽了屋顶平台和人工草坪。这不仅解决了城市用地紧张的问题，而且为高年级学生提供了游乐场地。

尼姆的设计理念表达了布鲁奈特和索尼尔如何成功地重新诠释一种在工业建筑中常见的看似老套的手法，即重复性的结构体系。在他们其他的作品中，如正在建设的勒阿弗尔(Le Havre)的音乐舞蹈学校和于1999年建成的Jussieu大学的临时建筑就体现了这种立面的设计精神。

在古镇尼姆，这个不落俗套的设计作品注定在既有的古老建筑氛围里掀起了波澜。由于这个建筑立面，许多评论家批评整个建筑结构太粗糙，太高科技化。

Another measure to reduce the dominance of the aluminium slats is a part of the facade itself. On the second and third floors, two large, loggia-like openings are punched through the entire depth of the new building to create open-air play areas here, as there was not enough room on the ground. Finally, the entire length of the new building is topped by a flying roof, a rigged, white canvas structure that shades the roof terrace and its artificial lawn. This is the play area for the older children, also installed above the city for lack of space elsewhere.

The design concept for Nîmes demonstrates how Brunet and Saunier succeed in using a homogeneous, seemingly banal, repetitive structural system that appears to be taken from industrial building, in a new and persuasive way. Other examples, such as the school of music and dance in Le Havre, under construction at the time of writing, or the temporary building for Jussieu University completed in 1999, illustrate this quality with comparable facade concepts.

In Nîmes, the attitude, uncompromising and yet not in any way dominant, is decisive in making the building a surprise in this historic place. Because of its facade, some critics

然而事实上，在围合和通透的相互影响下，它所采用的结构以及设计构成都沿袭了清晰的结构概念。建筑设计尺度合宜，在阳光的沐浴中，精致的铝合金百叶分格像闪烁在地中海天空下的白银。

小尺度的学校建筑和大尺度的罗马剧场从外部看都转化为雕塑的骨架构成。很自然，在开放空间中两个建筑物的体量不具有可比性。二者带来完全不同的空间感受。但纯粹从形式上看，二者都有室内外空间交融渗透的因素。剧场巨大的回廊向外部开放，而学校的铝合金幕墙将建筑向前延伸了一大步，使其连成一体，并与立面结合在一起。

第二次与布鲁奈特和索尼尔的作品相遇，作品中的后退式立面结构更好地阐述了他们的设计精神。我们离开位于塞文山脉(the Cevennes)和卡玛格岛(the Camargue)之间的尼姆市，来到圣日耳曼昂莱(Saint-Germain-en-Laye)，巴黎的西北郊区。在这里布鲁奈特和索尼尔设计建造了另外一个学校。将这两个设计理

have put down the entire structure as being too crude, too technical. In the interplay of closure and transparency, however, it is conclusively structured and all its parts follow a clear functional logic. There is no conflict of scale, and in sunlight the building and its many delicate slats gleam like precious silver under the Mediterranean sky.

The exteriors of both the school, on a small scale, and of the Roman theatre, on a large scale, are transformed into sculptural framework. Naturally, the buildings are not comparable in terms of their physical presence in framing open space. A very different impression results in each. But in a purely formal sense, in both cases the effect is that of the interplay of interior and exterior. The arena's large arcades open the galleries to the exterior, the school' s aluminium structure, extended far in front of the actual building, pulls together and structures the facade as a whole.

A second encounter will better illustrate how Brunet and Saunier master the work with reduced facade structures. We move from Nîmes, the city between the Cevennes and the Camargue, to Saint-Germain-en-Laye, a suburb north-west of Paris. Here Brunet and

▲ 圣日耳曼昂莱的国际学校

International School in Saint-Germain-en-Laye

念迥异的学校直接进行对比是一件很有趣的事情。国际学校拥有Château d' Hennemount的房产。学校原来的教学部位于城堡的底部，场地中其他的建筑也正在为学校所用。教学部位于两个两层的、有着人字形屋顶的短翼和一个连接翼中。建筑主体内包括学校特有的功能：一个报告厅，一个餐厅和一个图书馆。为了保持原有建筑群的风貌，布鲁奈特和索尼尔选择完全保留两翼并将新建的矩形建筑融入已有建筑中。

建筑师在这里选择了“无为”的设计思想，建筑场地位于一个公园中，被孤立的历史建筑所环绕，在这样的环境背景中，对历史建筑扩建的最佳解决方案就是不过分抢眼，保持谦虚的姿态，钢筋混凝土框架结构由修长的圆柱构成并隐现于外挂玻璃幕墙后方。外立面经过了特别的设计和处理。与尼姆的学校结构完

Saunier built another school. A direct comparison is interesting as the schools were realised at the same time, but based on entirely different concepts. The International School commanded a large estate with the Château d'Hennemont. The former working quarters located below the castle, as well as many other structures on the site, were already being used by the school. The working quarters consisted of two short two-storey wings with gable roofs and a lower connecting wing. The brief included some of the school's special functions: an auditorium, a restaurant and a library. In order to retain the character of the old ensemble, Brunct and Saunier chose not to touch the wings and to totally integrate the new rectangular building into the existing structure.

The architects refrained from doing too much. The special location, embedded in a park, surrounded by isolated historic buildings, and the direct extension to the remains of the existing structure in particular, called for a neutral solution with no need to flaunt itself. The architecture of the reinforced concrete skeleton with slender round columns is therefore emphasised only in the suspended glass skin. It is treated with particular attention and care.

全隐藏在铝合金百叶后面的处理方式相比，这里学校设计的立面结构是清晰可见的：自承重固定铝合金玻璃幕墙系统，完全是预制的、材质精良的构件与混凝土结构直接固定，这与1990年设计完成的多米尼克·佩罗(Dominique Perrault)的“Jean-Baptiste Berlier” 工业饭店 (Hôtel Industriel) 相似。杰罗梅·布鲁奈特和艾瑞克·索尼尔与佩罗有深厚的友谊。多年以前，他们在Vieille-du-Temple街开办了他们的事务所。在圣日耳曼昂莱建筑设计中他们采用了一种新的玻璃系统，即在两层玻璃之间增加一层遮阳百叶，这项技术上的创新成为该建筑作品中的亮点。它所采用的是固定的穿孔铝合金幕墙结构。经小孔的过滤，阳光变得分外柔和和纯净。走近窗户，透过小孔可以清楚地观察到室外的景观。这样一个特别的围护结构给人以意外的视觉效果。只有身处其中良久以后才会适应。建筑给人的最初感受完全不同于尼姆的学校，虽然在两个建筑中都采用了限制视线通过的遮阳构件。在这里不是两个完全分离的立面系统，而是将多层次结构整合在一起的节省

In contrast to the school in Nîmes, where the building structure is hidden by the slat wall, here the facade structure is immediately legible: it is a self-bearing fixed-panel window system with aluminium frames, completely prefabricated and made up of identical elements screwed to the concrete structure, similiar to Dominique Perrault's Hôtel Industriel “Jean-Baptiste Berlier” completed in 1990. Jérôme Brunet and Eric Saunier are friendly with Perrault. Years ago they took over his office space in Rue Vieille-du-Temple. What is new in the window system at Saint-Germain-en-Laye is the delicate sunscreen grating between the two layers of glass. It is a fixed perforated aluminium structure. It admits only subdued light because of its perforation, that texturally resembles cleaning tracks, but from close-up the view outside is hardly impaired. The ensuing character of a total enclosure is surprising. Only after some time does one grow accustomed to it, once seeing outside becomes possible from the right point of view. Seemingly different at first in terms of its organisation, the building does allow for a comparison to the school in Nîmes. In both cases, the architects chose a shading device that limits views into and out of the building. Here, however, this is

空间的系统。与尼姆学校相比，只在外窗中采用遮阳系统而忽略了其他围护结构有其难以避免的弊端，这使南面外墙在夏天需要采取额外的隔热措施。

圣日耳曼昂莱近郊的高尚社区中，建筑冷峻的方形体块，白天呈现镜面的外表，夜间则变成透明的立方体，这样的处理手法在教师中引起了争议。在天气好的时候，很多人感觉被封闭在建筑里，但是由于建筑的特殊地理位置——位于公园的中央，这样的处理方式可以被理解。但是，撇开建筑立面的光滑材质处理方法，布鲁奈特和索尼尔在建筑创作的道路上也开拓出新的天地。由此带来的争议也是在所难免的。

achieved not by two separate facades placed side by side, but by one compact space-saving system that integrates the different layers. In contrast to Nîmes, blocking direct sunlight only in the window element itself was a disadvantage. It resulted in excessive heating of the building's southern side and required the installation of a second shading device later.

The sober box in the upscale suburb of Saint-Germain-en-Laye, mirroring in daylight, transparent at night, is controversial among the teachers. In good weather, many feel locked up inside the building, which is understandable, located, as they are, in the middle of a park. But regardless of one's judgement of the smooth skin: with the facade, Brunet and Saunier are once again breaking new ground. Doing so inevitably entails fundamental discussions.

在尼姆学校建筑中所采用的装饰遮光百叶在圣日耳曼昂莱的国际学校中变为铝合金孔网，同样在巴黎的Jussieu大学中化身为立面外挂的玻璃格栅，由此可以得出：建筑师对材料的灵活运用会带来许多新奇的视觉效果。距离、视角和光影创造出丰富变幻的材质效果，给参观者时时带来不同的外观感受，有时甚至会产生错觉。

对建筑师以上两个作品的寻访引领我们走进了他们所营造的建筑世界。这两个学校建筑吸引着我们的目光，激发了我们走近并了解两位建筑师创作的热情。

The narrow slats at Nîmes, the delicate aluminium grating integrated into the glass in Saint-Germain-en-Laye, but also the angled industrial glass elements set in front of the facade at Jussieu University in Paris, make one thing clear: the architects are working on new variations of the visual effects that result from the specific use of materials. Proximity and distance, angle of vision and light create varying textures. The observer sees something different time and again, sometimes even believes in optical illusions.

Two visits marked the beginning of this reflection. They were chance encounters in different places. The school buildings caught the eye and stimulate interest in studying the work of the two architects more closely.

◄ 索恩河畔沙隆（Chalon-sur-Saône）
音乐学校的立面细部。

School of Music, Chalon-sur-Saône,
facade detail

▲ 位于巴黎Neuve Tolbiac街的住宅建筑

Residential building Rue Neuve Tolbiac in Paris

场所

对于杰罗梅·布鲁奈特和艾瑞克·索尼尔来说，文脉是建筑设计创作的基础。这不仅仅包括建筑所在地的自然地理环境和周围特有的环境特征，还包括一切与建筑相关的细节。建筑师基本的设计概念来自缜密的分析结果，通过分析得出全新的对现状空间的构思设计，这种构思只部分来源于传统的空间概念。但是最后，分析过程中所得到的“发现”往往对设计概念的发展全无影响。

对文脉的关注和研究是布鲁奈特和索尼尔设计理念中重要的组成部分。他们不热衷于生搬硬套某种成熟的建筑语言，或是制造标志性建筑语汇，来时刻昭示

Site

For Jérôme Brunet and Eric Saunier, context forms the basis of all considerations in a design project. This does not include only the built physical condition or the particular characteristics of the closer surroundings. These are fundamental considerations that deal with the particularities of the situation. The architects' basic design conception is developed from the results of the analysis. From the spirit of the site as found in a precise interpretation of the context emerges a new identity, an identity that is based only partially on the general spatial configuration. In the end, the "discoveries" often have no influence at all on the generation of the design concept.

Working with and within the context is important to Brunet and Saunier. They are not interested in unreflectedly forcing a specific, auto-referential architectural language on a

自身的存在性。事实上，他们更乐于创造出一件又一件个性鲜明、概念独特的作品，它的灵魂恰恰来自容纳建筑本体的环境，并与之结合得天衣无缝。但是，最初这种方式看上去是有缺陷的，因为一些鲜明的建筑手法不会反复采用而在其建筑设计中渐渐消亡，使得这两位建筑师的设计概念有专断之感。布鲁奈特和索尼尔相信建筑师不应该将建筑限制在抽象的美学范畴之内，而应带来太多的调和和中性的因素。因此，他们认为人们经常漠视建筑的现实存在性而更多的关注其他因素。对于他们来说建筑的场地和它与周围现实情况的关系才是决定性的因素。他们不是为了表达他们的设计意念而设计，针对每一次设计任务，他们所得出的解决方案的细节都因地制宜、相互迥异。

place. They are not concerned with a brand, an idiosyncrasy pushed on the other to unmistakably mark one's presence. Rather, they are concerned with developing, piece by piece, a unique and individual concept that originates in the place itself and is explicitly and unmistakably made for that place. At first, this may seem like a weakness, revealing a trait of arbitrariness in the architects' conceptual standpoint, because a signature clearly different from that of others appears to be missing in their many buildings and projects. Brunet and Saunier believe that architects should not limit building to its intellectual dimension, leaving too much tolerance and indifference. In doing so, in their opinion, one easily loses sight of the building's realisation and leaves it to others. To them, the building site and its connection to the reality of the place is decisive. They are not committed to building a manifesto of their thought. They are committed to facing the building task, arriving at solutions that are very different in detail every time.

这是他们坚定的创作立场，对此产生的建筑讨论尚未到来，虽然它可能很重要，毕竟这个时代需要偏袒性言论。在法国各个建筑学校中形成了不同的建筑学派阵营，他们主导着对建筑潮流发展的评论。布鲁奈特和索尼尔在20世纪70年代他们的学生时代里，正如他们所描述的，“被新现代主义所威慑”，但是他们现在有时候开始厌恶这个建筑学派，因为它不再具有创新活力，他们也拒绝成为“新柯布西耶追随者”。这个学派的设计手法在法国一些白色的、非常正式的建筑中大量应用而忽略建筑功能的需要。显然，对于布鲁奈特和索尼尔来说像让·努维尔这样的建筑师要比像亨利·戈丹这样的重要得多。

布鲁奈特和索尼尔对于他们要走的创作道路有着清晰的脉络。虽然他们的作品不断出现新的变化。在他们不断变化的建筑语言的背后，是一条在多个层面上永恒的坚持不懈的信念。工程的可操作性被摆在首位。在一个非正式的对话中，他

That a definitive “position” within the current, diffused architectural discussion is absent may be viewed critically, as times demand taking sides. The clearly articulated camps in the French architectural profession, formed over time in the context of the different architecture schools in Paris, dominate any discussion about current developments. Brunet and Saunier, in their student years in the seventies, were, as they put it, “terrorised by Neo-Modernism”. They have turned away from this school, which is no longer innovative, for some time now, and refuse to be “néo-corbuséens”. The reference is to the white, very formal architecture employed for every type of building regardless of its function, which continues to be of great relevance in France. Clearly, with this attitude, to Brunet and Saunier architects like Jean Nouvel are more important than architects in the like of Henri Gaudin.

Brunet and Saunier have good reasons for working the way they do. Although their handwriting continually comes along in new variations, it can be characterised upon closer inspection. Beneath the endless variations of their architectural language, one consistently finds a conceptual idea which proves convincing on several levels. What is practicable in a

们花时间来了解场地的详细情况，在这个过程中得出清晰的设计解决方案。到目前为止，在一个设计地点采纳过的创作思想从未在另外一个地方沿用或重复出现。每一个设计问题都有它相应的解决办法。结果决不是几何形式的堆砌，而是沿着布鲁奈特和索尼尔一贯坚持的根据功能和设计需要得出的“一种形式”。

为了更好地理解他们的工作方式，国立音乐舞蹈和戏剧艺术学院值得我们深入研究。这是一个小设计项目，是通过赢得1998年的设计竞赛而获得的。它建在法国南部的勒阿弗尔，就像法国在集权分散时期建造的其他建筑一样，这个过程持续了好多年。正因为如此，许多建筑师在这段时期越来越多地获得了由州和地区发起的建造学校类建筑的设计任务。

project is always at the forefront. In an informal dialogue, they take their time in getting to know the site. This process then leads to a clear design decision. To date, a concept realised in one place was never resumed or re-used in another context. Every design problem is seen in its own right. The result is never a conglomerate of geometrical forms, but always "one form" which Brunet and Saunier then consistently follow through to its realisation in terms of functional and design requirements.

In order to better understand the way they work, the Ecole nationale de musique, de danse et d'art dramatique merits a closer look. It is a small, but demanding project the architects were awarded after winning a competition in 1998. The ensemble is being built in Le Havre, northern France, as one of the many building projects realised in the course of the decentralization of France, a process that has been forced for years. For this reason, many architects are increasingly obtaining commissions to build institutions initiated by the state or by the strengthened region *en province*.

学校坐落于高密度的Cours de la République地区。场地位于原海港改造用地的边上，这里将修建新的学校设施。在法国的城市发展中，对原有海港的改建和再利用是一个非常重要的主题，它同样存在于南特和波尔多两个城市中。在勒阿弗尔，建校的目标不是为了建造一所为少数学生服务的封闭式学校，而是一所开放式音乐、舞蹈和戏剧学校，吸引和促进周围居民参与业余音乐和戏剧活动。为了解决相邻交通繁忙街道带来的噪声问题，布鲁奈特和索尼尔提出了一个特别的解决办法来实现建筑的开放特征。除了入口门厅，他们需要将立面大面积封闭起来。小的开窗是外立面仅有的开口。最初，立面从首层开始完全被不同形式的曲形穿孔铝合金玻璃幕墙外包，目的是为了遥望建筑，使幕墙具有标志性作用。但最后建筑师决定重新设计外立面，以垂直方向具有张力的金属拉索代替原有的“波浪”，而产生致密的纹理。拉索从底层飞跃至屋面边缘，使

The school is situated on the heavily-used Cours de la République. It is a site on the edge of a restructured section of the former harbour, which is to accomodate, among other uses, new university facilities. The restructuring and re-use of former harbour areas is an important theme in urban development in France, illustrated also by the cities of Nantes and Bordeaux. In Le Havre, the goal was to create not a closed-off institution for few pupils, but an open house of music, dance and theatre, to invite and motivate inhabitants to informally get involved with both music and theatre. The noise produced by the heavily-used road demanded that Brunet and Saunier develop a special solution to realise the desired open character of the building. Except for the entrance lobby, however, they felt forced to largely close the facade. Small window hatches are the only openings. Originally, the facade was to be completely enveloped above the ground floor by a second skin of curved aluminium perforated in different ways. The idea was to symbolise a stage curtain as a sign, visible from afar, of the new and open house. It was later decided to redesign the facade. Instead of “waves”, vertically tensioned steel cables will now create a dense texture. The cables span from above the low entrance floor to the edge of the roof, only

里面的建筑面若隐若现。该建筑外立面的拉索酷似竖琴的琴弦，使之成为了海港的标志，令人浮想联翩。这一设计创意成功地改变了传统的海港粗犷的城市形象。除此之外，拉索还可承载大型宣传板和灯光设施，面向城市宣传自己。对于这个简单的立面处理建筑师仍然考虑到了海湾的气候特征。精致的条板结构在这里就不是明智之举。新建筑蕴含着勒阿弗尔的未来：一方面，保留历史风貌，采用了“诠释”的手法，另一方面，对处于经济危机的海港城市来说它的转变至关重要，在这里也得到了完好的解决。

在二期建设中，演奏大厅悄然隐于已有的建筑群中。音乐学校的入口大厅将通往后面的演奏厅。学校将变为整体设计中的一部分，参观者在演奏厅内可以获知建筑内工作的进展情况。演奏厅的立面采用光滑材质的外墙涂料以及精致的金色铝合金挂板。

partially revealing the floor levels behind. This is to make a symbolic reference to the harbour, while also creating associations to musical strings, for instance those of a harp. The idea successfully reinterprets the rough character of the site marked by shipping and navigation. In addition, the system allows for the hanging of large panels as well as projections and light installations to advertise the events inside toward the street. But with the simple facade, the architects also took into account the raw climate of the Channel. A delicate structure of slats would have been the wrong choice here. The new building is to embody the future of Le Havre: on the one hand, its history is preserved, and yet in a sort of "translation", its changes, of great meaning for a harbour town in economic crisis, are also illustrated.

In a second building phase the auditorium is to be added in an unpretentious way and with limited means. The music school with its lobby will then lead into the auditorium toward the back. The school will become part of a whole, and the visitors in the auditorium will learn about the work going on in the building. The unspectacular facades are to be clad in smooth, subtly golden aluminium panels.

◄ 勒阿弗尔音乐学校的立面原型

Facade prototype of the music school in Le Havre

布鲁奈特和索尼尔在几年前曾接受一个类似的设计任务。到目前为止，在第戎(Dijon)南部的索恩河畔沙隆，这个设计作品还散发着迷人的魅力。音乐学校位于圣科姆(Saint-Cosme)的中部，这是一个新兴地区，位于索恩河，火车轨道和公路之间，主要由2层小型建筑和20世纪60和70年代的建筑构成。新建学校在这里有一个核心的功能，即成为小镇新兴中心的焦点，因此该项目的筹划备受瞩目。建筑师量体裁衣地在场地中引入了一个中央林荫走廊，它穿过场地和建筑，室外的台阶将建筑的中心部位和走廊联系在一起。与勒阿弗尔的项目不同的是，为表达建筑的使用性质，这里的演奏大厅在建筑中处于举足轻重的地位。它

Brunet and Saunier had faced a similar design problem a few years earlier. And yet, in Chalon-sur-Saône, south of Dijon, a strikingly different approach is obvious. The music centre is located in the middle of Saint-Cosme, a recent development between the Saône river, railway tracks and a peripheral road. Small two-storey buildings, but also housing slabs of the sixties and seventies are scattered side by side on the site. The school has a central function here, it is to become the focal point of the small town' s new quarter, that is being planned with great enthusiasm. The architects developed a concept tailored to the task by laying out a central pedestrian path that traverses both the terrain and the building, hoping that the building' s glazed middle area, accessible across a flat ramp of stairs, be frequently open to pedestrians. Different from the project in Le Havre, here the concert hall was deliberately weighted in order to express the building's function. As the seat of the philharmonic orchestra of Chalon-Bourgogne, it is of importance beyond the region. The hall takes the shape of a round volume, clad in light birch wood, that was inserted into a glass lobby and stands out from the rest of the building above the roof line as

还作为沙隆布戈涅爱乐乐团的演奏排练场地，因此建筑的重要性大大超过了作为地区性的音乐演出场地。演出场地是一个圆形的大厅，外饰配以浅色白桦木，一端插入到玻璃门厅中，另一端突出建筑高出天际线。大厅后部的舞台外观装饰着曲面红棕色板材，像一件做工考究的“家具”。与演奏厅、门厅和通道相邻的是学校建筑群。在外部它们各自独立，但是它们的外表形象都与音乐相关，在这样重要的地理位置上凸显建筑的整体性，像一个符号系统。分散的开窗又或是像冲压机的打孔带，看似自由分布却与装饰在外立面的、随混凝土和铜质金属板交错变化的线条相一致。水平向的窄条，使我们想到熟悉的建筑师最基本的设计概念，它出现于布鲁奈特和索尼尔早期的作品中，例如在圣克莱尔市(Hérouville-Saint-Clair)的警察局建筑侧立面设计中。提契诺(Ticino)建筑的影响在当时占了主导地位。

well. The area to the back of the hall with the stage is clad in bent, reddish-brown wood panels and placed like a distinctive “piece of furniture” in the middle of the neighbourhood. Beside the passage, the auditorium and the lobby, there is another adjoining building volume, the slab structure housing the school. It is clearly set apart in its exterior, but in order not to be too understated in this prominent location, it was given a skin that makes references to music. It suggests a notational system or the punched tape of a grinding organ, as scattered openings, seemingly placed at random, align with the stripes that result from the alteration of concrete and copper plate cladding. With its narrow horizontal bands, the facade should be seen as one of a series of facades all similar in their basic concept, realised by Brunet and Saunier early on in their collaboration, for instance in the lateral walls of the police department of Hérouville-Saint-Clair. The influence of Ticino architecture at that time is obvious.

▲ 圣克莱尔市警察局

Police department in Hérouville-Saint-Clair

在索恩河畔沙隆，建筑师准确地实现了建筑三个基本功能：向市民开放；使演奏大厅成为小镇标志性建筑，通过圆形的玻璃大厅欢迎市民的来访；为在演奏厅周围活动的人们，如在学校的人们，提供远景，从外形上表达建筑的特殊功能。

为了更好地全面理解布鲁奈特和索尼尔两位建筑师的设计思想，我们还需研究另外一个清晰表达建筑师完全不同的设计方法的建筑——蒙彼利埃(Montpellier)的法兰西银行。这个设计任务是建造一个区域性的银行分理处，建设地点靠近城市中心，周围环绕着优雅的建筑和茂密的树林。业主对工程预算有限制。令人吃惊的是，这里所选择的方案清楚地展现出别人曾经用过的建筑手法。布鲁奈特和索尼尔辩解说他们选用的解决方案完全是源自于对设计任务恰当的理解。他们在这里所运用的设计手法被证明是通过实现每一个设计细节

In Chalon-sur-Saône, the architects precisely articulated the three required functions of the building: to be open to the town; to mark the concert hall as a new sign recognised from afar, drawing in visitors through the encircling glass lobby; and to keep passers-by at distance in certain parts such as the school, without forgetting to express their specific functions in the facade.

To obtain a better idea of Brunet and Saunier's range of design, it is important to study a building that manifests yet another, entirely different approach: the Banque de France in Montpellier. The task was to create a regional headquarters of money that would be representative in its location, a part of town close to the inner city, characterised by elegant buildings and dense trees. It goes without saying that financial means were not a problem. Astonishingly, the architecture chosen here clearly displays an existing architectural position and would seem to have been designed by others. Brunet and Saunier defend the selected "style" by sustaining that it resulted from the consistent interpretation of the building task. Their use of the available means proves that here an idea was pursued and

▲ 昂热(Angers)办公楼建筑设计

Project for an office building in Angers

而使之完善升华，而在其他工程中对细节的推敲则不可能达到如此精湛的程度。在蒙彼利埃，路德维希·密斯·凡·德·罗设计的巴塞罗那展览馆建筑成为他们创作的样板：没有比例，没有功能，围护墙体和地面浮现于流动的、纪念性的和严肃的空间氛围中。密斯·凡·德·罗的作品对两位建筑师产生了深远影响，特别体现在罗宰昂布里(Rozay-en-Brie)的初中建筑中。在蒙彼利埃银行中，优雅、独立的外墙挂板构成了银行大厦外表的基本结构。不向公众开放的相邻建筑则具有完全不同的独立的气质，其玻璃表皮在构成变化和其他细节上经过了实践的验证，并应用于后来的工程，如巴黎Neuve Tolbiac 街住宅，昂热办公楼建筑(1999年的获奖作品)等。

followed through up to the realisation of the details, while in other projects it had not been possible to allot similar attention to detailing. In Montpellier, Ludwig Mies van der Rohe' s Barcelona Pavilion was the model: not in scale, not in terms of function, but in the basic conception of the wall and floor slabs with the flowing, celebratory and stately space. The degree to which Mies van der Rohe's work serves as an example to the architects is apparent in other projects also – particularily in the secondary school in Rozay-en-Brie. In the Montpellier bank, the elegant, clearly legible and apparently isolated wall slabs create the basic structure of the banking hall. The adjacent building, not accessible to the public, has an entirely different, independent quality. Its glass skin has since been employed as a well-tried facade composition–in variations and with other detail solutions – in subsequent projects like the residential building on Rue Neuve Tolbiac in Paris, or the office building project for Angers, a competition prize of 1999.

► 罗宰昂布里的初中学校

The secondary school of Rozay-en-Brie

昂热可以被视为建筑师建筑生涯中一次独特的尝试，在这里涌现了一个全新的办公建筑的设计概念，它沿用了近几年流行的“开放空间”的概念。建筑立面被解读为一个非实质性的“保护壳”。而真正的办公空间插在类似框架结构的空间中。建筑的结构系统可以被看作是“房中房”，在这里双重结构系统构成了整个建筑体量。每一部分都很灵活，有一些甚至突出于墙面，可以随时被重组。

Angers should be noted as an exceptional design in the architects' œuvre. A new concept for office construction emerges here, pursuing the endeavours of recent decades on topics like the "open office". The facade is to be read as an immaterial "protective envelope". While the actual offices are inserted into a frame-like structure. The system can be seen as a type of "house-in-house" where individual duplex units make up the building volume. The parts are flexible, some projecting slightly from the facade, and can be recombined at any time.

◄ 巴黎弗兰德街住宅的立面细部

Facade detail of residential block Rue de Flandre in Paris

▶ 巴黎Neuve Tolbiac 街住宅的立面细部

Facade detail of residential building Rue Neuve Tolbiac in Paris

另一个建筑的场地位于巴黎第19区的弗兰德街(Rue de Flandre)，这是一个颇富争议的区域。20世纪60年代，这条笔直的马路在北侧被拓宽，在此期间，大量的19世纪的古建筑被毁，取而代之的是高层公寓，但只有部分是沿着新的街道的。在马路的南侧，就是新建建筑的用地，已建成了几个大的建筑，它们退到街道的一边，处在现有的建筑之间。由于体量上的不均衡，街道呈现出不寻常的城市景观。对于站在街角的建筑，布鲁奈特和索尼尔试图使建筑获得平衡。他们尝试了一种对巴黎近郊建筑类型的新解。在20世纪80年代的晚些时候，阿尔多·罗西(Aldo Rossi)已经开展过巴黎古典居住建筑的再研究。当时，他被委托在靠近拉维莱特(La Villette)的弗兰德街设计一个住宅建筑，紧邻克里斯蒂安·包赞巴克(Christian de Portzamparc)设计的音乐学校。罗西所设计的学校的特征是高耸的双重斜坡的屋顶和位于底层的密柱式开敞画廊。根据建筑师的想法，建筑应唤回里沃利街(Rue de Rivoli)时代，因此，建筑将“对城市产生的影响力要远大于它自身的前卫性”。现在我们可以看到这种概念是失败的。建筑丝毫未表现出类型学的发展，在理论上它保持了十分有争议的建筑形象。在设计上，该住宅毫无论证价值，它的拱廊也没有任何实际使用价值。

布鲁奈特和索尼尔设计的住宅建筑则大不相同，朝向弗兰德街的立面给人以深刻的高雅的气息。在这里建筑对环境的呼应不是孤立的，而是与周围紧密联系，建筑不是一个标志物站在这里向我们传递信息，而是环境纹理有机的组成部分。诚然，大气的立面设计没有贯穿于楼层的所有部分。

The building where site is best discussed, but also in the most contradictory terms, is located on Rue de Flandre in the 19th arrondissement of Paris. In the sixties, the road, dead straight, was widened on its north side, in the course of which numerous turn-of-the-century buildings were demolished and replaced by high housing blocks which only partially follow the new street line. On the south side of the road, where the new building is located, a few big blocks were added, set back from the street between existing buildings. As a result of the ensuing disparities in scale, the street has assumed unusual urban form. With their corner building, Brunet and Saunier tried to balance this structure. They attempted a new translation of the typology of the Parisian Faubourg building. In the late eighties, Aldo Rossi had already made it his theme to reinterpret the classical Parisian house. At the time, he was commissioned to design a residential building near Rue de Flandre in La Villette, next to Christian de Portzamparc's music school. Rossi's building is characterized by high mansard roofs and an open gallery with tightly positioned columns on the ground floor. According to the architect, it should recall Rue de Rivoli, and in so doing, the building is to a have "a greater effect on the city than buildings of the avantgarde". The concept was a failure, as we know today. The building shows no typological progression, and perseveres in a theoretical, very questionable image. In plan, the flats have no demonstrable qualities, and the arcades are of no real use.

This is different in Brunet and Saunier's residential building. The street facade on Rue de Flandre is characterised by impressive elegance. Here the reaction to the site is not an isolated gesture, but refers directly to the neighbourhood. The building is not a symbol, placed there to send us a message, but becomes an integral part of the texture. Admittedly, the largesse and clarity of the facade is not carried through to the organisation of the

从室内看，建筑流露出奥斯曼式(Haussmann)的风格。不论如何，在后面，在这个难以处理的场地最密集的部分，紧邻高架铁路，并在含混的灯光环境下，建筑师大胆采用了新的设计方法营造出最合宜的设计作品。

这个方法不是为了纯粹的试验，因为它还要接受审视的目光。在这个设计问题上没有试验的余地。尽管如此，布鲁奈特和索尼尔有解决问题的勇气。到目前为止，相比较而言，他们在住宅设计领域里所获得的最大的成功就是Neuve Tolbiac 街住宅。受罗兰·施威策尔(Roland Schweitzer)所做的城市总体规划的控制，场地几乎没有发挥的自由。这里将成为“建筑的展览”：不同的住宅组团——部分由弗朗西斯·索莱尔(Francis Solar)和菲利普·加佐(Philippe Gazeau)设计——将环绕在多米尼克·佩罗(Dominique Perrault)设计的法国国家图书馆周围，立面再一次被赋以宏伟的设计外观。玻璃建筑是对20世纪60、70年代的追忆，那是一个建筑师乐于复古20世纪20、30年代建筑的时代。

flats. Seen from the inside, the references to Haussmann seem forced. Nevertheless, the project dared to go new paths – and in the back, in the tightest part on this difficult site, next to the old viaduct, with problematic lighting conditions, the architects achieved best possible work.

This approach is not based on pure joy of experimentation, which would have to be viewed critically. There is no room for experimentation in this kind of design problem. Brunet and Saunier, however, do have the tenacity to engage in a given problem. And yet, in comparison, their great success in housing is the new residential building on Rue Neuve Tolbiac. The site allowed for little liberty, as the urban development was already laid down in a masterplan by Roland Schweitzer. It was to be a “building exhibition”: different residential blocks, among others by Francis Soler and Philippe Gazeau, are to surround Dominique Perrault's Bibliothèque Nationale de France. Once again, the building is captivating in the largesse of the facades. The glass building is reminiscent of the sixties and seventies, a period that the architects like to refer to, more so than to the twenties and thirties.

◄ 简·阿昂尔(Jean Anouilh)和皮埃尔·卡德罗丝·拉克斯(Pierre Choderlos de Laclos)设计的用丝网印制的彩色文字用在Neuve Tolbiac街住宅阳台的栏板装饰上

Serigraphic fragments of texts by Jean Anouilh and Pierre Choderlos de Laclos were applied on the balcony parapets of the residential building Rue Neuve Tolbiac

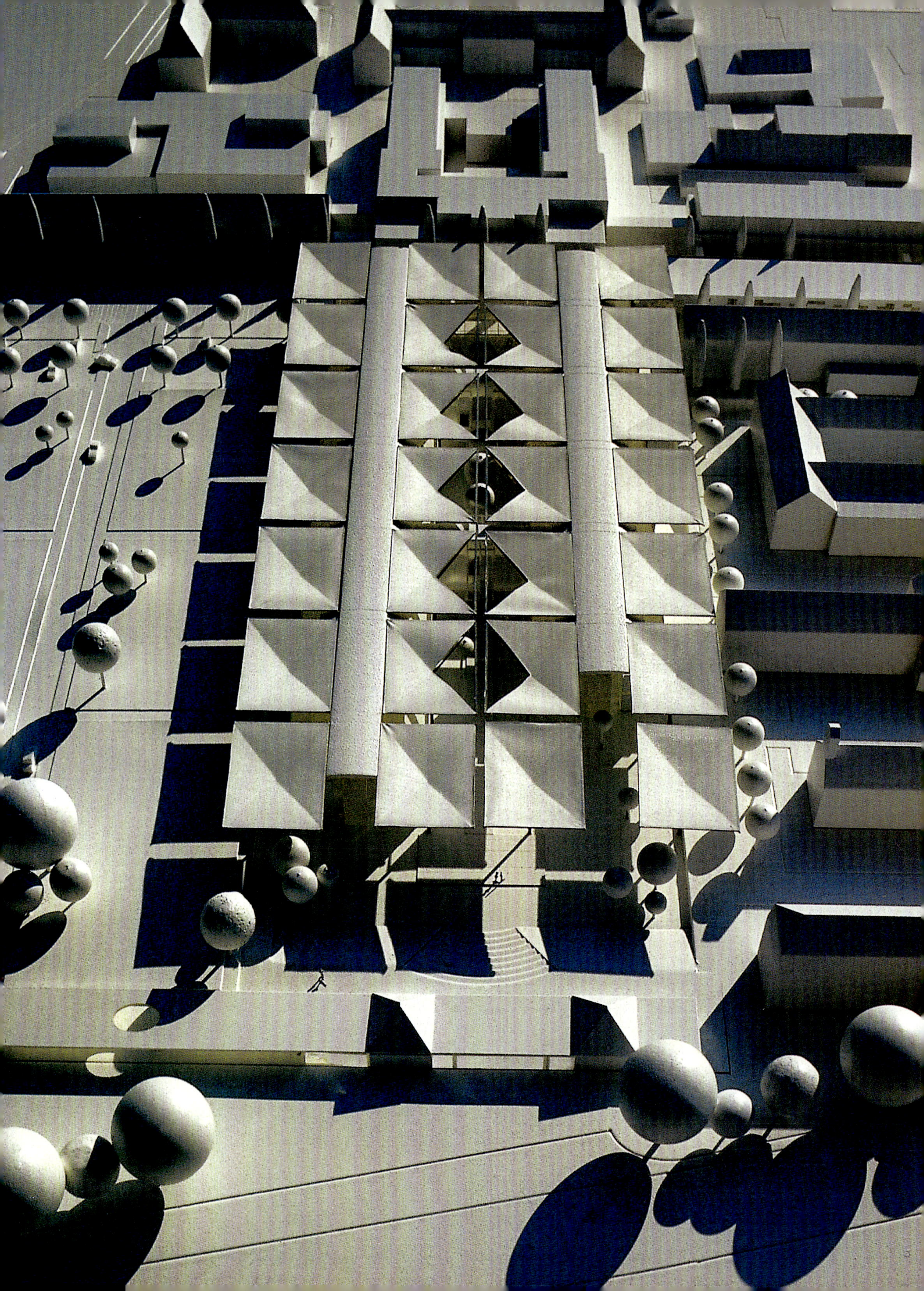

布鲁奈特和索尼尔对场地的成功把握还体现在两个医院建筑中。卡昂(Caen)和图尔(Tours)的医院项目是法国医院建筑革新的典范。为了更好地整合，新的医院同样靠近市中心。特别是卡昂的儿科医院不再是患者的“收容所”，而是一个为公众服务的开敞空间。建筑向南北分别延伸至宽阔的乔治-克莱蒙丝尔大街、一所历史悠久的公园和一个现存的妇产医院。新建的建筑平面组织呈十字形，可以从不同的方向进入该建筑。在卡昂，年均降雨天数是184天。屋面悬挑遮蔽室外空间是当地建筑的特色。这里还将围合一个供患者和来访亲朋见面的内院，将医院变成孩子们的乐园。同时这里还规划了一个室外剧场。室外悬挑屋顶由28个方的薄膜结构组成。两个长建筑的立面分别由一个巨型钢结构框架构成。每一个框架网格上都外挂着由卡昂当地特产的轻质石材或玻璃制成的扁板条格子。为了营造出丰富的图面，玻璃窗将根据儿童和成人的不同视高而改变。

In what different ways Brunet and Saunier deal with site is also demonstrated by two hospital projects. The projects for Caen and Tours are examples of a changed philosophy of clinic construction in France. New clinics are once again being located close to the inner city in order to better integrate them. The pediatric hospital in Caen, in particular, is no longer to be a refuge for “outcasts”, but a largely open place of public passage. The building extends north-south between the wide Avenue Georges-Clémenceau and the remains of an old park and an existing birth clinic. The new building is organised in cross shape, permitting access from all sides. In Caen, the average number of annual rain days is 184. Large roofs covering the exterior areas are a special feature. There will be a patio where patients can play with visiting friends, turning the hospital into a children' s house. An open-air theatre has also been planned. Outdoor roofing comprises 28 square umbrellas with textile membranes. The facades of the two long building fronts each consist of one enormous steel frame. Each frame borders a grid of flat panels made either of light stone from the Caen region or of glass. Windows shift from children's to adults' eye-height in order to create a vivid picture.

◀ 卡昂的大学医院

Project for the university hospital of Caen

▲ 尼姆的税务局

Project for the tax office in Nîmes

图尔的医院项目处在市中心。位于一条狭窄街道两侧的两幢建筑构成了现有的医院建筑，一幢建于20世纪初，另外一幢建于20世纪60年代。新规划为儿科门诊的建筑同样临街。作为群体的第三个建筑，它一开始就展露出完全不同的外部形象，具有和谐美好的气质。一层由强调水平方向的玻璃幕墙围护而成，上部是轻质的石材饰面。立面吸收了图尔城蜿蜒古街的地貌特征，但又不是简单地模仿，它主要由外窗的分格构架而成。上部一条连续性的窄幅玻璃作为轻型的围护结构。这样可以有效地减少建筑体量对环境的冲击。

场地是设计的主题。在近几年中，在城市的中心，在第三行政区，在位于共和国广场和蓬皮杜中心之间的Dupetit-Thouars街都可以找到杰罗梅·布鲁奈特和艾瑞克·索尼尔特别的工作场所。他们的办公室设在一座建于世纪之初，正立面拥有大面积玻璃窗和坚实柱子的华丽建筑中。巴黎的美景像陈列在橱窗中的画卷，从

The project for Tours is located in the city centre. Two buildings on both sides of a narrow street currently make up the hospital. One building dates to the turn of the century, the other to the sixties. The new building planned for the pediatric clinic also fronts the street. As a third building in the ensemble, it appears at first to have a clearly different facade. And yet its exterior is designed to create a harmonious integration. The ground floor area is glazed, emphasising the horizontal orientation. A closed volume clad in light stone rests above it. The facade picks up on the typology of the old winding street of Tours without trivialising or turning into a pastiche. Window slits structure the facade. Above, a continuous narrow band of glass serves as a light-weight closure. This ingeniously reduces the thrust of the block.

Site is the topic. For a few years now the special work site of Jérôme Brunet and Eric Saunier can be found in Rue Dupetit-Thouars between Place de la République and Centre Pompidou, in the 3rd arrondissement, in the heart of the city. Their office is located in a wonderful turn-of-the-century building with large window fronts and cast-iron columns. The

▲ 国立高等机械与航空学院

Ecole nationale Supérieure du Mécanique et d'Aéronautique – Futuroscope in Poitiers

角窗流进室内：右侧是一个古老的拥有完美比例的砖石结构的学校，左侧是一个以落地窗和锌皮屋面为特征的小型建筑群。这个区域里的建筑是这个城市的见证，它形象丰富，能激发设计者的创作激情并从这里研究出新的建筑材料——即使布鲁奈特和索尼尔真正关注的是20世纪50、60年代的建筑。布鲁奈特和索尼尔在工作室中的合作实现了完美的分工。在设计阶段，两个人像训练有素的团队，之间的交流和协作以大量的草图形式实现。其实，在成立工作室之前他们就已经这样做了。他们共同为皮埃尔·路易斯·弗罗斯工作了3年。在那里他们掌握了这样的工作方法。

view from the corner window is a true showcase of Paris: to the right is an old brick school with a well-proportioned facade, to the left are small old buildings with French windows and zinc roofs, typical for the area. This site is surrounded by references to their city, a place rich in images, a place for inspiration by what you see and for developing new material from it–even if Brunet and Saunier's true passion is the architecture of the fifties and sixties. Brunet and Saunier as a studio means the ideal division of labour in a partnership. In the design process, the two are like a well-trained team, passing balls in form of innumerable sketches. They were doing so even prior to founding their own office. For three years, they both worked for Pierre-Louis Faloci. He showed them the way and they are endebted to him.

▲ 从帕拉瓦莱弗洛 (Palavas-les-Flots) 的康复中心疗养泳池向外面的海滩眺望

Looking out to the beach from the rehabilitation clinic medicinal baths in Palavas-les-Flots

平衡

杰罗梅·布鲁奈特和艾瑞克·索尼尔不认为他们自己和建筑之间有“手足之谊”。他们不想成为某些特殊“法则”的“忠实信徒”，这从他们作品的外部形象可以看出来。其次，与他们的谈话往往集中在他们最喜欢的话题上，即反对建筑中的教条和生硬的线条。他们希望能够自由地处理少数重复性的元素以及尝试用不同的方法解决不同的问题，但这并不意味着求异。设计的目标是保持创作的自由度以获得充分的灵活性。这听上去容易，但是在实际操作中却充满挑战性。这样大致可以保证在概念设计阶段，他们在创作中所激发的火花不被抹煞。

Balance

Jérôme Brunet and Eric Saunier do not consider themselves to belong to a “brotherhood” of architecture. They do not want to be “servants of faith” of a particular “order, characterised by its unmistakable outer appearance. And again, the conversation with them focuses on their favourite topic: the uncompromising rejection of a dogmatic, definitive line in their architecture. They want to be able to deal freely with the few recurring elements and solve every problem with a new approach, while not treading paths that are too unconventional. The declared goal is to retain a measure of freedom in order to offer sufficient flexibility. This may sound easy, but it is also challenging. It probably guarantees that in the conceptual phase, the playful aspect of their work is not neglected.

整体考虑他们的作品，仔细研究设计过程中草图所体现的意念，在这个具有典型特征的区域中，新建筑的立面设计具有异乎寻常的重要性。方案中的概念设计以及空间关系都与立面的设计息息相关。这并不意味着“表皮”可以取代本质的东西——这样的想法将会错失建筑师创作的关键点。通常，空间的组织不像设计概念之于外立面那样凭想象和创造性地得出。这在圣日耳曼昂的学校和帕拉瓦莱弗洛的儿童康复中心的建筑设计中分别有着明显的体现。

在住宅设计中，到目前为止此类的建设项目在巴黎正逐渐减少，它立面设计和平面设计经常会存在着相互矛盾。在Neuve Tolbiac街的95户单元住宅楼外观极具优雅气质，占据整个建筑进深的入口空间亦使来访者由衷惊叹。但与之形成鲜明对比的是，上部的楼层通道却狭窄和黑暗。很明显，建筑的外立面构思以玻璃幕墙为主体，正因为如此，使一些楼层空间存在缺陷。在占据整个建筑进深的单元，

Considering the œuvre as a whole and studying the ideas sketches produced in the design process, it is striking that next to the reference to site, the facade design is unusually important. Conceptual considerations in plan or spatial relationships are always thought of in terms of their relation to the facade. This is not to say that the “clothing” supersedes what is essential–such an assumption would miss the point of the architects'work. But the spatial organisation does not usually seem as imaginatively or ingeniousely worked out as the conception for the facade. This is especially evident in examples such as the school in Saint-Germain-en-Laye or the rehabilitation clinic for children in Palavas-les-Flots.

In housing construction, which to date has all been subsidised projects in Paris, there is even a certain discrepancy between the facade and the flats. The building volume on Rue Neuve Tolbiac with its 95 units is extremely elegant and draws visitors into surprisingly spatious entrance lobbies, open across the entire building depth. The upper stories, however, are characterised by narrow, dark interior hallways. Clearly, the building was conceived for the glass skin. Because of this, some flats are at a disadvantage. Those that span the depth of

有的部分被迫于一角，例如，居住区通过大型的推拉门直接和外部环绕的阳台相通，控制着周围的一切。因为在住宅设计中存在限制，因此在这里不可能实现创新型设计理念。然而，在这里采用玻璃幕墙的表面印上了简·安昂埃尔和皮埃尔·考德拉斯·拉考斯的篇章片段手法的痕迹，这是个很吸引人的主意。

在弗兰德街，立面和内部空间质量的差异表现得更加明显。杰罗梅·布鲁奈特和艾瑞克·索尼尔选择了一个几乎可以贴上“古典”标签的立面形式，这不同于他们的设计风格。但是大型落地窗和白色推拉百叶窗不只是立面的一部分，它们连成排，本身就是立面，形成了一幅变幻多样的画面。它们之间仅通过水平向的楼板线条分割。从外部看，它再一次打破了传统印象中单元住宅的外部形象，并使建筑融入了古街一侧的街景中。与该地区早些时候建造的住宅项目相比，新

the building, for example, are somewhat cornered in parts – the living areas, extending directly on to the surrounding balconies thanks to the large sliding doors, dominate everything else. Because of the restricting brief it was not possible to achieve here an innovative conception for housing construction was not achieved. Rather, the priority was the attractive idea of having a glass facade across from the Bibliothèque Nationale de France, with serigraphically applied fragments of texts by Jean Anouilh and Pierre Choderlos de Laclos.

In Rue de Flandre, the discrepancy between the facade and the spatial quality of the flats is even more obvious. Brunet and Saunier opted for a facade that could almost be labeled “classical”, unusual for their practice. But the large French windows with their white sliding shutters are not only a part of the facade: added in long rows, they are themselves the facade, creating a diversified picture. They are separated only by the narrow horizontal bands of the floor slabs. From the outside, once again, the image of social housing has been dispelled. The building fits into the old street-front of a side street, but compared to earlier

► 帕拉瓦莱弗洛儿童康复中心立面细部

Facade detail of the rehabilitation clinic for children in Palavas-les-Flots

▲ 通向帕拉瓦莱弗洛的康复中心入口的棚架

Pergola at the entrance to the rehabilitation clinic

建住宅就像一个奢华的公寓，具有与众不同的气质。当然这只限于建筑的外部特征，而内部空间却显得相对复杂。外部表露的信息与内部完全不同。建筑师设计了一个穿越不同楼层的玻璃窗。建筑的外立面清楚地表达了它内部具有连续性的楼层，但它并没有与实际内部的楼层结构高度相对应。位于落地玻璃窗上部的近半米高的结构高差用深色镜面反射玻璃封闭，但是从外部几乎看不出来。立面上的玻璃条带用来作为栏杆。因为栏杆的高度必须为1m，所以在玻璃窗的外侧必须增加一层玻璃。尽管如此，真正的玻璃窗高度仍然有2m，从地（楼）面到顶棚的高度是2.60m。

social housing programs in the area, it resembles luxury condominiums and is clearly set apart. This concerns only the building' s exterior. The interior is a more complicated matter. For what appears on the outside turns out to be very different on the inside. The architects designed a window that crosses floor levels. The continuous floor levels so strongly articulated on the outside therefore do not actually correspond to the height of the structural floor slabs behind. The difference of a bit less than half a metre means that the upper zone of every French window, covering the actual floor behind, is made of dark mirrored glass, but hardly read as such from the outside. The band in the facade must therefore be interpreted as the parapet. Because a parapet must be one metre high, an additional pane of glass had to be added in front of the windows. Nevertheless, the actual window is still two metres high, and the floor to ceiling height is 2.60 metres.

在佛兰德街，第一次出现了一些与布鲁奈特和索尼尔作品相陌生的东西。建筑的立面与建筑的内部空间不相一致，但完美的比例与周围环境的共生，使我们毫不置疑建筑的品质。同样，在近郊的建筑实践中也可以有创造激动人心的建筑外观的机会。但是，他们的设计理念留下了些许疑惑，在这里他们首次尝试了将建筑立面设计成为“表皮”，并且使它仅具有外观形象的作用。这与他们一贯坚持的创作思想出现了明显的分歧。布鲁奈特和索尼尔的立场就是不想被冠以某种标签，他们针对每一个新的设计任务都重新设置一个假定，就像建筑师第一次移动的方向与他的手所指的方向并不一致一样。例如在Neuve Tolbiac街，二层以上建筑几乎没有交通空间，户型紧凑，内部空间组织简单并遵循了住宅设计的常规导则。与平面分离开来，重新设计组织外立面在这里是最佳的解决方法。两个顶层单元甚至被设计成复式住宅。

In Rue de Flandre, for the first time something occurs that is foreign to the work of Brunet and Saunier. The facade claims to be something that does not match the interior. The qualities of the building, well-proportioned and well-adapted to the ensemble in Rue de Flandre, are unquestioned. Also, the idea of working with the Faubourg-era building type may certainly create exciting possibilities in facade design. But their concept leaves one puzzled. Here they have, for the first time, designed a “skin” that in a sense does not want to go beyond being a show facade. This is a departure from the position they usually defend with such clarity. Brunet and Saunier’ s position that they do not want to be categorised and that they begin every design problem on a new premise leaves one at a loss here, as for the first time the architects have moved in a direction that does not match their hand. As in Rue Neuve Tolbiac, little space remained for circulation on the upper floors. The flats themselves are tight, their spatial organisation is simple and they conform to social housing guidelines. The best possible solutions were achieved by rearranging the available surface areas. The two top floors were even made into duplex flats.

布鲁奈特和索尼尔的建筑语言从不被潮流所影响，即使在佛兰德街设计中，但是在他们的建筑语言中有一个发展趋势，即将立面作为一个重要的转换因素。清晰的立面结构、精细的表皮纹理在整个设计中扮演着重要的角色。在一些项目中，比如勒阿弗尔的音乐中心，立面处于如此重要的地位，以至于真正的结构和建筑内部空间的组织不再被关注。理由很明显：建筑欲重新展现自身。建筑师们更愿意赋予他们所设计的作品以抽象的内涵，将其用被誉为“表皮”的介质包裹起来。这层“表皮”不仅仅与实际结构相脱离，而且在一些实例中，它们直接呈现在建筑前面。一些人看到尼姆学校的铝合金条框构成的外立面以后，可能会产生一种冲动，就是拿出大剪刀来剪断它们，把里面学习的孩子们“解放”出来；或是产生拆除帕拉瓦莱弗洛的康复中心底层的部分木条的想法。但是，这些都是错误的判断。在“表皮”设计概念的背后隐藏着一种设计技术：抽象的表皮可以营造出比例优美、外表光鲜、风格出俗的建筑。同时它还可以作为建筑的过滤器和保护层。

Within Brunet and Saunier’s range of architectural expression–never influenced by fashionable trends, not even in the case of Rue de Flandre–there is a general tendency to use the facade as an important element of transformation. A clearly dominant facade structure, more precisely a facade texture, is of great significance to the overall design. In some projects, the music centre in Le Havre for instance, the facade is so dominant that the actual structure or organisation of the building can no longer be made out. The reasons are well-known: the building is to represent itself. The architects prefer to lend their buildings an abstract quality, to supply them with a neutral layer they like to refer to as *enveloppe*. This layer as “cover” is not only distinct from the actual structure, in many examples it is in fact set in front of the building. Some people may feel inclined to get out their wire clippers to “liberate” the school children from behind the aluminium facade in Nîmes, or may want to remove some of the wooden slats from the base of the rehabilitation clinic in Palavas-les-Flots, but this would be wrong. Behind the idea of the *enveloppes* there is a design technique: the abstract skin creates a subtely proportioned building, gives it a face and makes it unique. The cover is both filter and protection.

虽然“双层皮”对品质无损，但它同样有其自身的问题。它将人们熟悉的建筑形象分解或是简化。这在帕拉瓦莱弗洛项目中，位于海滩的底层部分上有清楚的表达。带有非凡气质的水平方向的木条屏蔽了外部的视线，使内部的药物浴疗和健身房的活动不受干扰。在Jussieu，大学校园位于巴黎西南，双层皮的应用处于较基础的水平。在这些临时的建筑中，竖向的工业玻璃元素用螺栓固定成简单的排列组成外立面。建筑师再次成功地创造出令人兴奋的具有冲击性的视觉效果，但是，走近仔细观察发现，设计立面的底部，建筑本体并没有显露出来，建筑的外层皮是可见的，而内层皮以及它们之间所夹的空间却很难分辨出来。此时，参观者的头脑里会闪现一系列的问题：内外表皮之间的空间里会发生什么？这里蕴含了怎样的特别内容？一般情况下，两层之间只有60cm宽，里面如何打扫呢？但是一旦退到远处，人们会被建筑的整体精彩的外立面设计所吸引

Without prejudice to its qualities, the “double shell” also has its problems. It creates a certain dissolution or a reduction of the observers' accustomed image of a building. In Palavas-les-Flots, this is apparent in the powerful basestorey on the beach. The horizontal wooden slats, formally joined in an extraordinarily elegant manner, ensure that from the outside one hardly catches a glimpse of what is going on inside the medicinal baths and the gymnastics rooms behind. In Jussieu, the university campus in southwestern Paris, the double shell was realized on a far more basic level. In these temporary buildings, the vertical industrial glass elements were simply set in front of the facade and screwed in. The architects again succeeded in creating an exciting visual stimulus, but the downside of the design is revealed, just as in Palavas-les-Flots, when seen up close: the building itself is not really present. The outer facades are visible, while the inner facade and interspace of the building behind can barely be made out. What happens to this space in between? What are the specific qualities of such a zone? How is the space, often up to sixty centimetres wide, cleaned, for example? These questions come to mind, but are quickly forgotten, when, at a

◀ 巴黎Jussieu大学校园建筑的立面细部

Facade of the university building on the Jussieu campus in Paris

而将刚才的问题抛到九霄云外。Jussieu的立面设计如此之奇妙，在于从不同的角度和变幻的光影环境中观察都会油然而生神秘的色彩。

人们不禁疑惑，这样的立面材质设计概念会将我们带向何方，永恒这个要素在建筑中很重要，但在这里却缺失了，同样的疑虑也出现在其他的一些项目设计中。没有什么看上去是有机的，或是厚重的。那些属于建筑的稳定的特性，在这里也荡然无存。视觉上的稳定不是布鲁奈特和索尼尔设计的主旨。但是当对整体立面加以审视以后，其优雅性和复杂性将前面的不良感受一扫而光。并不只是一个展示出了强有力的姿态立面或是一种复杂的、高科技的、不锈钢质感的结构形式，这样精细的处理方式赋予了建筑激昂的活力。布鲁奈特和索尼尔从不在设计中添加肤浅的符号语言或是结构主义的元素。立面的每一个局部都相互协调，没

distance, one becomes aware of the building' s overall structure as a successful composition. The sight of Jussieu, very different depending on point of view and light of day, can be almost mystical.

One may wonder where this textural conception of facades will take us. The impression may occur that the momentum of permanence, so important in architecture, is missing here, generating doubts as to some of the buildings and projects. Nothing seems organic or massive. These qualities, intrinsic to building and lending it stability, are missing. Visual stability is not the subject of Brunet and Saunier. But even this certain discomfort is dispelled at the latest once the homogeneous structure of a facade is exposed in all its elegance and rich complexity. Not one of the facades needs a powerful gesture or a complicated, high-tech, stainless steel construction with its fake fragility to lend the building a strung-up dynamic touch. Brunet and Saunier do without superficially emblematic or structuralist elements. No part of the facades is tensioned, no detail is artfully crafted for

◄ 圣日耳曼昂莱国际学校的立面细部

Facade detail of the International School
in Saint-Germain-en-Laye

B
C
D

有一个细部是完全从美学角度出发的。所有构成元素都不是新发明创造，它们都源自工作中所涉及的已有材料。木支架、铝合金条和一切使用玻璃的地方——在圣日耳曼昂莱市政厅内使用的玻璃柱，卢佛尔宫工作间的玻璃屋面，总之，将普通材料通过特殊的设计和组合来实现预期的构造效果。

这种非教条的、直接的，但决不是单一的、可转换的设计方式正代表了杰罗梅·布鲁奈特和艾瑞克·索尼尔的设计理念。这也是他们变得如此忙碌的关键。多样性、个性以及某些概念上的矛盾性被融入他们的作品之中。由此，在对他们的作品解读之后，我们逐渐原谅了在帕拉瓦莱弗洛项目中暴露在外的木质推拉门所发出的声响，以及小心地处理从海滩吹来散落在地板缝中的细砂。

purely aesthetic reasons. None of the formal elements are inventions. They all result from the work with existing materials. Wooden planks, aluminium slats and above all glass – glass columns in the Saint-Germain-en-Laye town hall, the glass roof of the Louvre workshops – are used in a way that the desired effect is created only by the particular treatment and joining of the materials.

This undogmatic and direct manner, never monotonous or exchangeable, characterises the work of Jérôme Brunet and Eric Saunier. This is what makes them so engaging. Heterogeneity, individuality and a certain measure of conceptual contradiction seem to be an integral part of their work. And one feels moved to forgive the sound of the wooden sliding doors, which form the most exposed facade layer in Palavas-les-Flots, forever grinding slightly in the floor tracks filled with sand blown over from the beach.

◄ 圣日耳曼昂莱市政厅的中央大厅

The central hall of the town hall in Saint-Germain-en-Laye

小学，尼姆

小学距古城的中心——古罗马斗兽场仅有250m远。这是一个由金属和玻璃构成的建筑，与尼姆地区的典型建筑类型相异，它表面由铝合金装饰条构成，当刮北风时这些装饰条波动得很厉害。

原来的“La Placette”学校建于20世纪初。在扩建过程中，原有建筑除了南立面之外其余完全被保留。新建筑完全被安排在原建筑的前面。新建筑的平面组织异常简单。新旧建筑之间设置了一个罩有玻璃屋面的走廊。一排原来属于旧建筑南立面柱廊的柱子被保留下来并矗立在走廊的中间，但是它们已经失去了他们原来所扮演的结构作用。受到来自于现有建筑场地和想保留南向广场要求的限制，新建筑别无选择，只能沿着已有的校园建设。因此带来了集约化的建筑平面设计，房间被布置在走廊的两侧。

新建筑的钢筋混凝土结构与原有的建筑高度相同，因此，新建筑内的教室空间高度和现代建筑标准相比异常高大，开阔。新建筑的南向和东向的玻璃幕墙立面隐藏在遮阳百叶后面，带来了高级气派的感觉，百叶是由铝合金条构成，距玻璃幕墙60cm。对于建筑师来说，立面是一个巨幅的软百叶窗。每个单元的遮阳百叶都可以沿中轴旋转，创造了丰富的光影效果。只有栏杆是固定的。

在东侧，一个高塔状的体量里面设有楼梯、电梯以及管理员办公室和图书馆。在底层平面，标高略低于室外街道标高，设有多功能观众厅。教室设在三层平面以上。部分屋面铺设了人工草坪，成为高年级学生的活动场地，他

Elementary School, Nîmes

The elementary school is located only 250 metres from the Roman arena in the heart of the old town. It is a steel and glass building, untypical for its site in Nîmes, with an aluminium slat skin that vibrates heavily once the Mistral begins to blow.

The original “La Placette” school was built at the turn of the century. In the course of the extension, the old building was left untouched with the exception of its south facade. The new building has literally pushed itself in front of the old one. In its organisation, the building is extremely simple. An internal passage covered by a glass roof was inserted between the old and the new building parts. A small row of cast-iron columns, once part of the colonnade of the south facade, was preserved, but located now in the middle of the new corridor it is no longer structurally relevant. Given the limited dimensions of the site and the constraints resulting from the need to retain the open square to the south, there was no alternative but to build directly along the side of the existing school. On the inside, this resulted in an economical floor plan, with the rooms arranged on both sides of the corridors. The new, reinforced concrete structure is the same height as the old building, creating unusually high classrooms for today’s standards.

The new south and east facades are made up of a glass wall hidden behind a sunscreen-curtain that has a very technoid feeling to it, composed of aluminium slats and set 60 cm in front of the glass wall. To the architects, the facade is a transcription of a huge Venetian blind. The shading elements can be pivoted on a central axis, creating a rich play of shadows. Only the parapets are fixed.

To the east, a tower-like block containing stairs, lifts, the caretaker’s flat and the library was added. The ground floor, slightly lower than street level, includes a large multi-purpose

► 从
瓦尔
L'Ora

View
the r
the T

▲ 白色金属拉膜结构覆盖下的屋面空间被开发成为高年级孩子们的课间活动场地

The break area for older children is located on the roof under a rigged white canvas structure

▲ 在第二层设置了宽阔开敞的室外走廊

The "schoolyard" is a large loggia on the buildings second floor

► 建筑的西立面内设疏散楼梯与原有建筑相邻

The western front side with fire-escape and existing building to the left

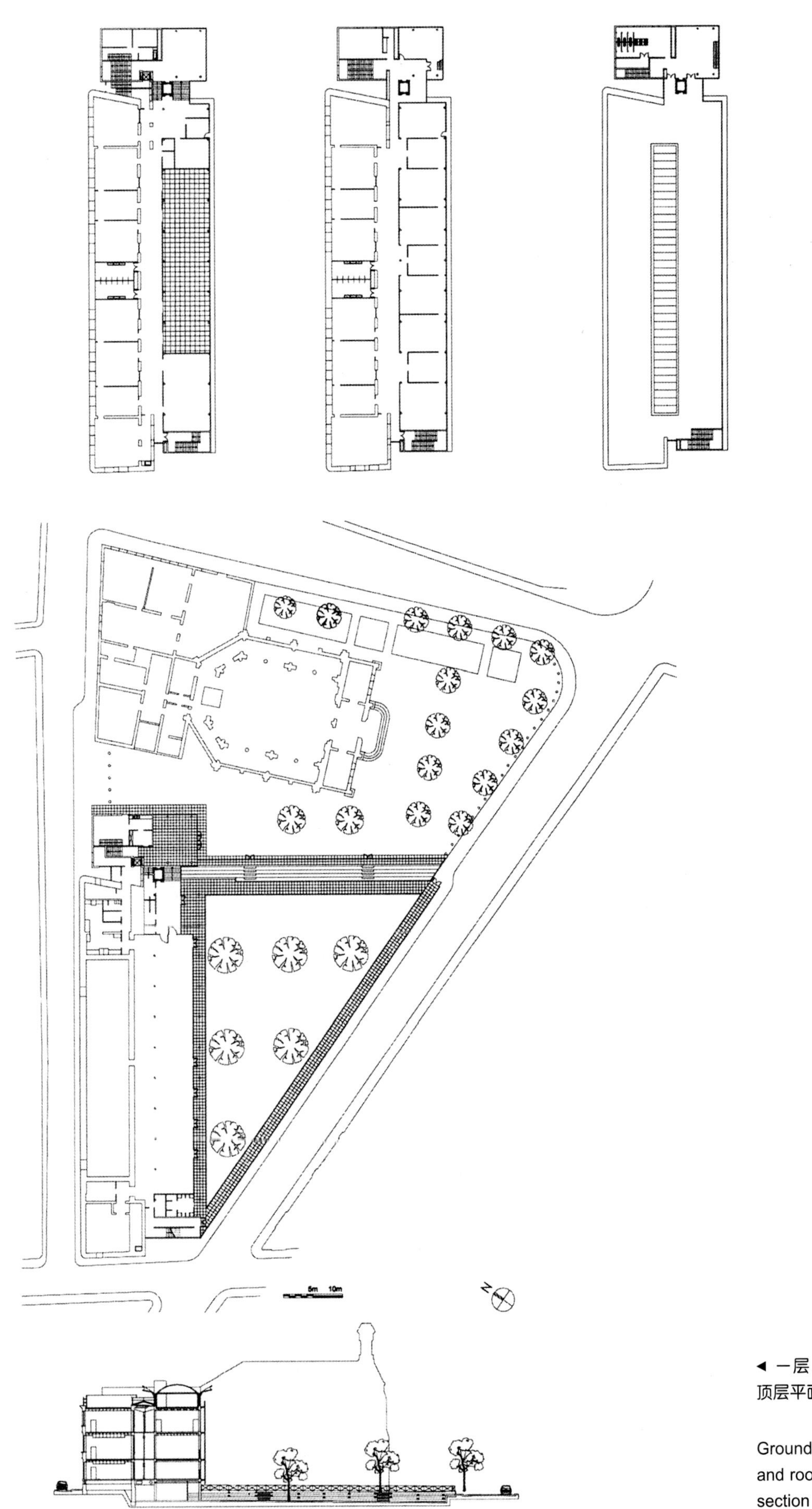

◄ 一层、二层、四层和顶层平面图和剖面图

Ground floor, first, third and roof floor plans and section

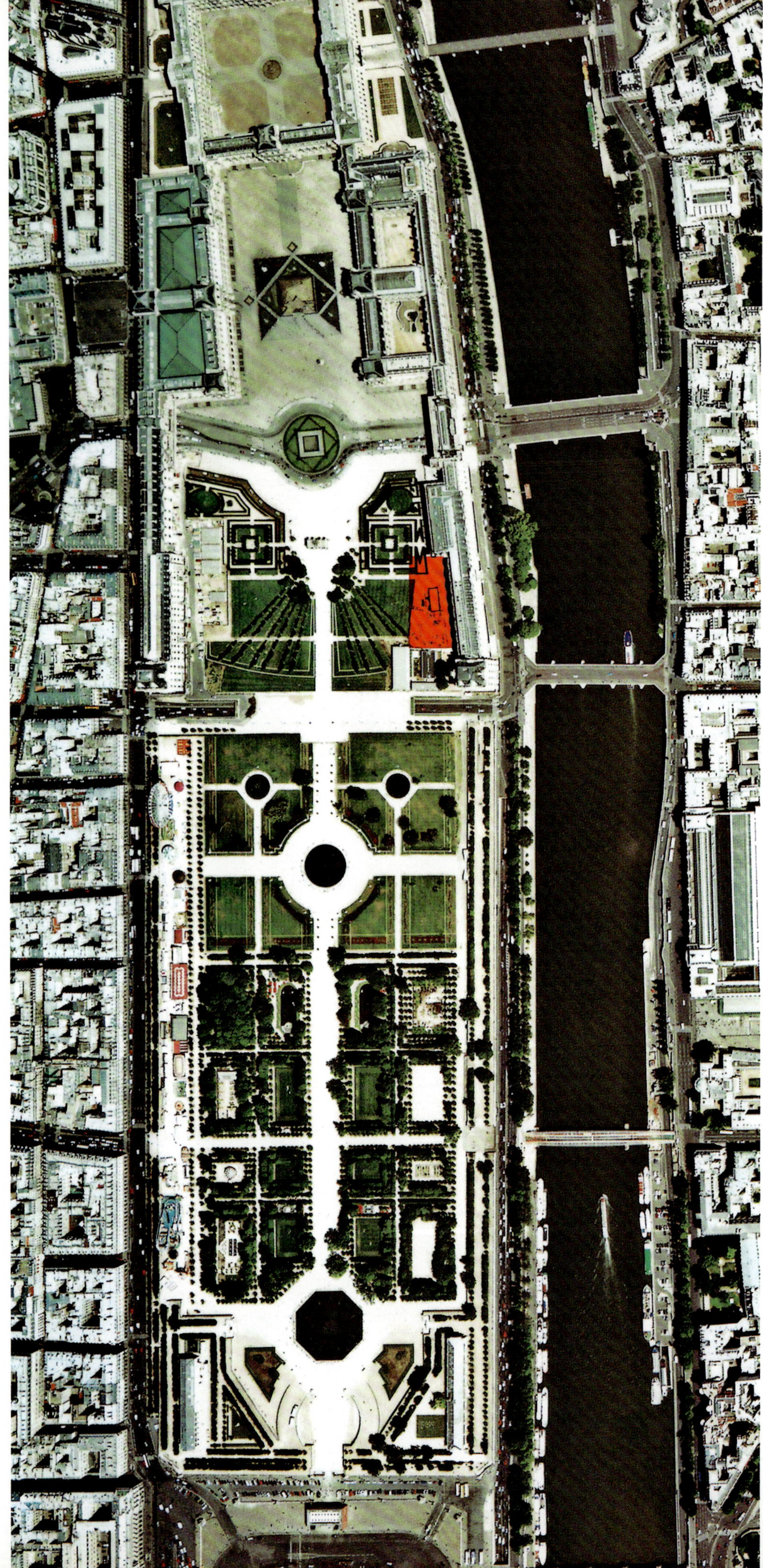

卢佛尔宫研究实验室，巴黎

Louvre Research Laboratories, Paris

在过去的十五年中，包括著名建筑师贝聿明在内的大量建筑师曾经被委托在卢佛尔宫的不同地方设计新建筑或进行建筑改造。布鲁奈特和索尼尔的设计作品几乎处于完全被隐藏的状态。这个设计任务是将在1931年建立的保存和记载艺术作品的研究实验室搬迁到一个新的、宽敞的、3层的地下室里，这个任务很困难，而且如果设计不好就会身败名裂。每年大约1500件艺术品在这里由科学家和来自法国博物馆的修复工作者进行检测和分类。设计中会有多大的回旋余地呢？建筑师们知道这些艺术专家原来的工作环境是在位于卢佛尔宫南端的Pavillon de Flore的顶层，对外有绝佳的视野，可以欣赏到卢佛尔宫和城市的美景。现在将他们的工作室搬移至大约地下12m处，即在Jardin du Caroussel的下方，他们的工作环境将变得完全不同。

在这里，建筑师的设计概念很简单，连续性的玻璃幕墙贯穿地下3层空间形成一个中庭。在两侧平面之间架设有连廊作为交通联系。中庭面宽只有4m，在Jardin du Caroussel的标高层上，覆以16m长的玻璃屋面。其由单独垂直的玻璃元素构成，它们像棱镜一样对透进来的光线起到折射作用。玻璃屋面的坡度为3%，走到这里，就像是走在绿树环绕的水面。

建筑中庭的光线不足以满足每个房间的采光要求，但是它在建筑内部和天空之间建立了一种重要的视觉联系。为了满足各个工作区域最大的通透性，中庭

Over the past fifteen years, numerous architects besides Ieoh Ming Pei have been commissioned with new buildings and building alterations in various parts of the Grand Louvre. Brunet and Saunier' s project remains almost entirely hidden. The difficult, compromised task consisted in relocating research laboratories for the conservation and documentation of objects of art, founded in 1931, to a large, new, three-level basement. Approximately 1500 pieces of art are examined and classified here each year by scientists and restorers of the French museums. What leeway was there? The architects were well aware that the experts' previous workplace had been at roof level of the Pavilion de Flore at the southern end of the Louvre, commanding spectacular views of the Louvre and the city. These people were about to be relocated to spaces up to 12 metres below ground beneath the Jardin du Caroussel. They had to be presented with something quite unusual.

The concept is simple. A continuous glass skin connects the three subterranean levels across their entire height to a central atrium. Bridges on the upper floors connect the two sides. The hall-like space, only 4 metres wide, is covered by a glass roof 16 metres long at the level of the Jardin du Caroussel. It is made up of separate vertical glass elements that project the captured light downward like prisms. The glass roof slopes at 3% and to the pedestrian looks like a reflecting surface of water framed by a planted border.

The central play of light is not enough to light the rooms, but it establishes the crucial visual connection to the sky. The central space with its spot-fixed glass walls, which demanded great transparency within the different parts of the workshops, has created a new type of

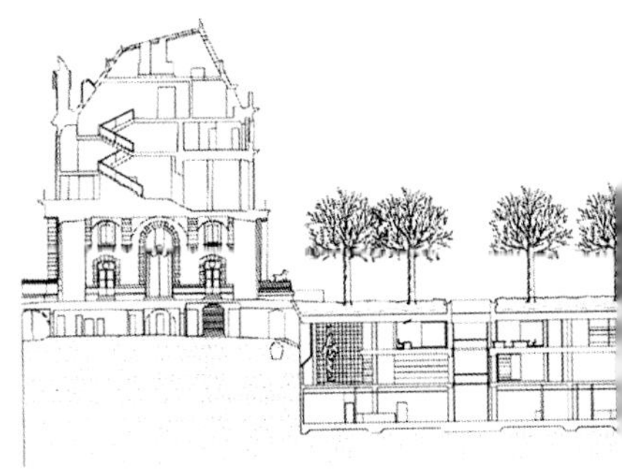

◄ 在Jardin du Caroussel的下方，大卢佛尔宫南翼的实验研究室剖面

Section through the research laboratories below the Jardin du Caroussel at the south wing of the Grand Louvre

▶ 实验室三层完全依靠中庭采光（左）

The three levels of laboratories are lit exclusively by a central skylight

▶ 中厅上方的玻璃屋面是由玻璃构造节点自承重的（右）

The glass roof above the courtyard is self-bearing with its glass joists

由点式玻璃幕墙围合而成，营造出一种全新的空间特质，人们被强大的空间感吸引，所以会忘却因缺失窗外的景观所带来的遗憾。对于建筑师来说，结果是具有隐喻性的，从某种意义上讲，这里的空间具有宗教气氛，每一条从黑暗走向光明的通道都弥漫着神秘的气息。

投射进来的光线没有遇到任何障碍，屋顶和墙面，包括支撑结构没有一处不是由玻璃制成的。屋面是由4.60m长，0.6m高，4层各1.5cm厚的玻璃肋支撑的。墙体被设计成为结构的表皮。玻璃在这里不仅是填充构件，还转化为综合性的结构元素——这种使用方式在法国还是第一次。

spatial quality, characterized by spaciousness that almost makes one forget the absent views. To the architects, the result is a metaphorical, in some ways a religious place, where the path from darkness to light has a mysterious quality.

Entering light encounters no obstruction. Neither the glass roof nor the workshop walls include structural elements not made of glass. The roof is self-supported by the glass elements that are 4.60 m long and 60 cm high, consist of four layers of glass, each 1.5 cm thick, and function as joists. Walls are designed as structural skins. Glass is not simply inserted. It is transformed into an integral structural element–used in this way for the first time in France.

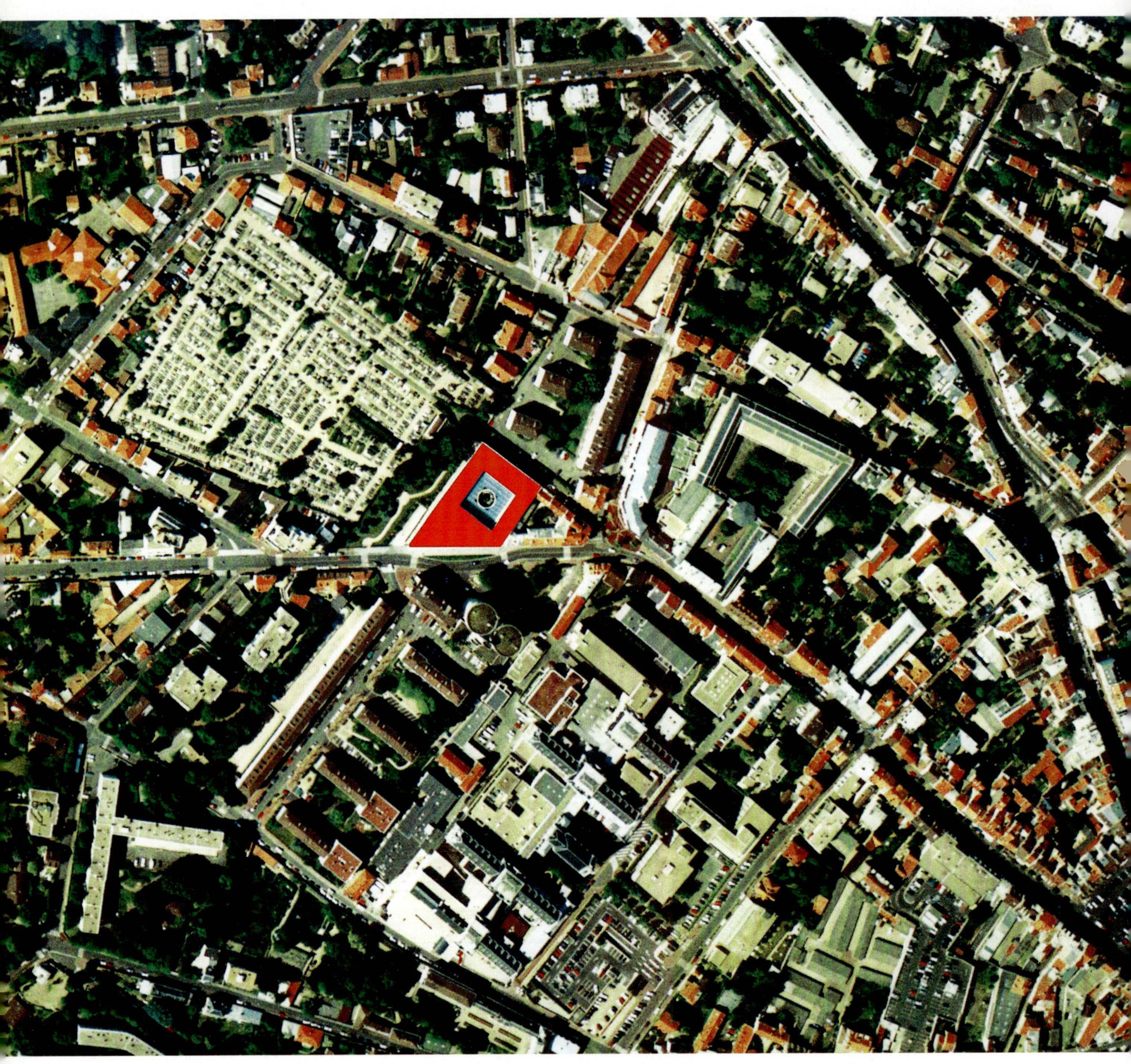

市政厅，圣日耳曼昂莱

圣日耳曼昂莱是位于巴黎西北部的一个知名的绿色地区，这里的市政管理部门分布在几个不同的建筑中显得不合时宜。为了将这些分散的机构整合起来，当时的市长麦克尔·派瑞卡特提出要在城市的中心地段建设一个新的市政厅。在竞赛投标之后，布鲁奈特和索尼尔获得了设计委托。

该建筑环境周围不是临空的，而是繁华的Léon-Désoyer大街沿街立面的延续。这个方盒状外形建筑的两个相邻立面形成了鲜明的对比。沿街立面被非常高的隔离物分开，上面两层的玻璃外墙向内深深地凹进去。尽管它不俗的地位，但新建建筑与周围的环境融合得非常好。在对着隔壁墓地的一端墙上悬挂着城市的徽章。在一层，深隐在竖向分隔后面的玻璃幕墙形成一个内凹的弧线使之成为一个令来访者备感亲切的入口空间。

与之相对比的是相邻的另外一个立面，一个从一层到三层完全由玻璃幕墙构成的非常长的立面，它朝向墓地和一个公园。

当来到市政厅后，来访者首先走进门厅，这里设有问讯处，开放式楼梯，玻璃景观电梯以及通向办公室的走廊。穿过布置在建筑中央轴线上的电梯厅一路走来，来访者被引入一个巨大的方形大厅。这里是市政厅的工作人员的接待处，他们的桌椅都布置在离墙稍远的位置上。在中央，是一个圆形中庭，里面种有植物，庭

Town Hall, Saint-Germain-en-Laye

In Saint-Germain-en-Laye, a respectable green suburb north-west of Paris, the municipal administration was inappropriately housed in several buildings. In order to pull the different departments together, Michel Péricard, mayor at the time, commissioned a new building on a central site. Brunet and Saunier were awarded the commission as the result of a competition.

The building is not free-standing, but part of the street-front development along the busy Rue Léon-Désoyer. The two facades of the block-like building are strongly contrasted. To the street, the facade is divided by full-height partitions, with the glass front of the two upper storeys set far behind. Despite the grand gesture, the building is well-adapted to the modest neighbouring structures. Only at the building's distinctive vertex that faces the adjoining cemetery, an unbroken wall surface sports the town emblem. On the ground floor, the glass facade behind the dividing walls wraps inward in a long, concave curve to create an inviting entrance area.

In contrast, the second, very long facade, oriented to the cemetery and a public park, is an enormous glass front, uniform on all three levels.

Upon entering the building, visitors reach the lobby with the information desk, open staircase, glass lift and access to the side corridors with the offices. Along the side of the lift, positioned in the building's central axis, one is led into the large, square hall. Municipal employees greet visitors at desks placed at a distance from the walls. In the middle, the

▲ 建筑有两个对比强烈的立面，一个沿街，另外一个朝向隔壁的墓地

The building has two distinct sides: one to the street, one to the adjoining cemetery

院的周围由向室内倾斜的玻璃围合而成，形成了一个倒圆台的空间。这个别致的空间最大限度地接纳外来的阳光，并且成为上两层方形空间的对景。

人流密集的接待大厅最特别的地方就是极为轻盈的玻璃屋面，为了防眩光，所以在玻璃屋面的下方衬了一层帆布。包括侧边的附属部分在内，整个750m^{2}建筑平面的围护结构和结构钢梁完全由十字交叉断面的玻璃柱支撑。每一个柱子都是由三层玻璃构成，共厚2.5cm。中间一层厚1.5cm并起到结构的承力作用。每个柱子的截面尺寸为22cm×22cm，4m高，可承受6吨的荷载。由于没有防火保护层，这里需要对这种新型结构作详细的检验试验。与建筑师合作的工程师是马克·马里诺夫斯基。基于这个概念，建筑师成功的营造了一个拥有开放的大空间构架和纤细富有张力的屋顶结构的接待大厅。

space opens toward a round atrium with plants, separated by curved glass that tilts inward. This open space admits ample light and grants views of the simple facades of the two upper floors that frame the central courtyard on four sides.

A special feature of the highly frequented hall is the extremely lightweight glass roof with the sun sail above. Besides the lateral attachments, the entire 750 m^2 surface and its steel beam structure is supported by twelve glass columns with a cross section, fixed at both ends by steel sheet. Each column is made up of three glass plates with a total thickness of 2.5 cm. The middle layer is 1.5 cm thick and has the load-bearing function. In section, a column measures 22 cm by 22 cm, is 4 m high, and can take a load of 6 tons. Not only for reasons of missing fire-proofing, extensive testing was necessary to have this hitherto unique structure approved. It was developed in collaboration with engineer Marc Malinowsky. With this concept, the architects succeeded in creating an open and spacious central hall with a roof substructure of strikingly slender dimensions.

◀ 沿街立面被有力的竖向线条分隔，入口处向后延伸形成一个内凹的弧线

The street facade is characterized by powerful vertical divisions and an entrance area that swings inward in a concave curve

▼ 朝向墓地的立面局部

Partial view of the facade facing the cemetery

► 接待大厅的玻璃屋面和它中央圆形的室外庭院被上部两层的办公室用房所环绕

The square roof of the hall with its central atrium is surrounded by two floors of offices

► 接待大厅完全由玻璃屋顶覆盖，下部由12根玻璃柱子支撑

The central hall is entirely covered by a glass roof supported by twelve cross-shaped glass columns

底层及上面隔层的平面图和沿街的立面图

Plans of ground and upper floors and elevation on Rue Léon-Désoyer

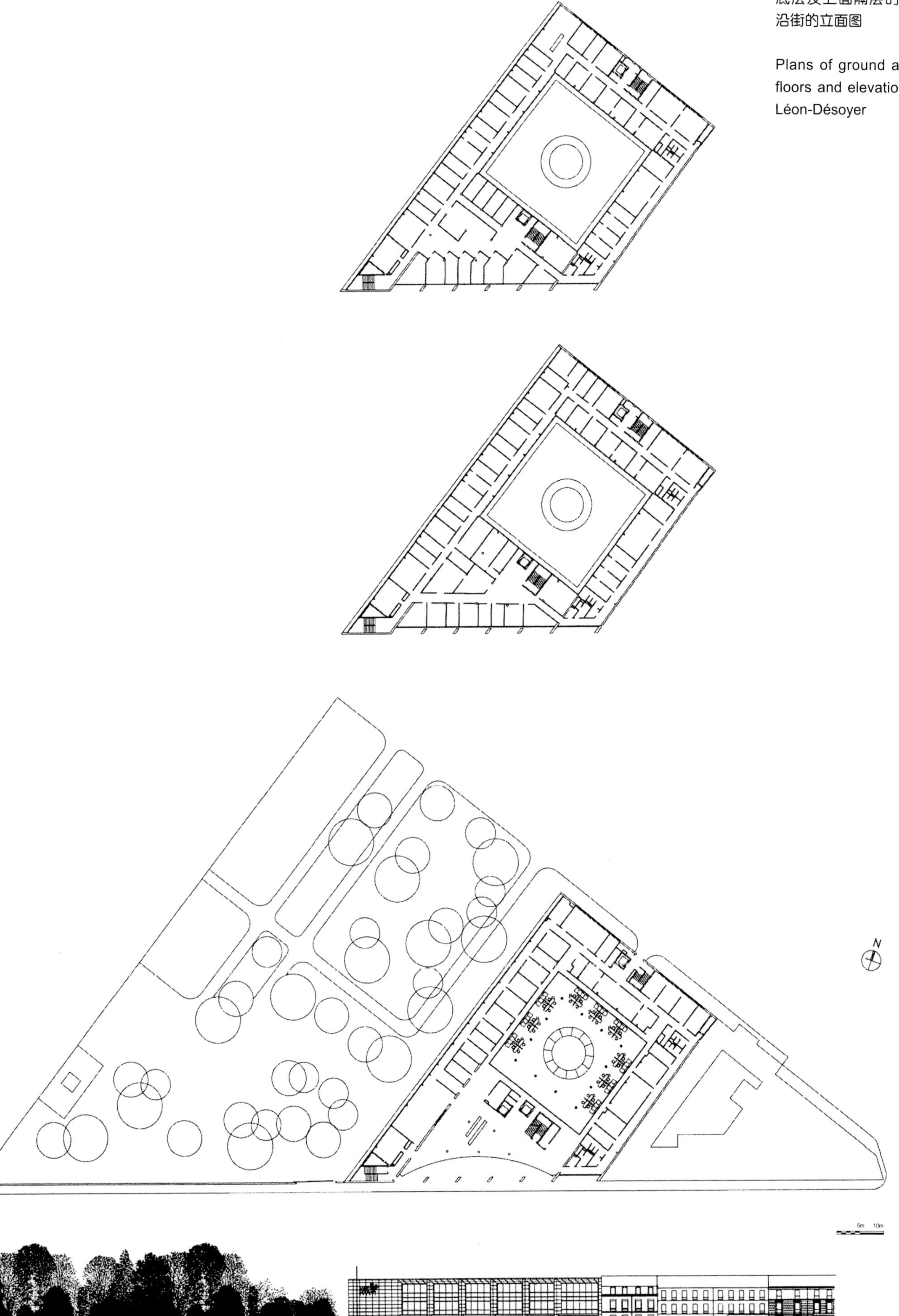

中学，罗宰昂布里

位于塞纳马恩省的罗宰昂布里中学具有很高的声誉，它位于一个新建住宅区和一个超市之间。冷静易识别的外表使这所建筑从周围的环境中脱颖而出。在它的“城市空间”中，建筑组织清晰，层次表达分明，比如：围合的广场可以作为中央广场和开放式的环廊。从类型学上讲，这与该地区的大面积的农场环境特征是相关联的。只是，在这里这些农场的封闭性被广场的通透开放性所代替。

主建筑是一个低层的方盒子，它的基本组织就像是一个回廊。一条轴向的通道由南至北贯穿学校并一直延伸到室外。到目前为止，只完成了建筑的一期工程，因此学校建筑目前呈U形，建筑师的设计概念尚无实现。

整个玻璃立面呈现出独特的气质，建筑通透性甚至可以穿透巨大的室内庭院。建筑师一直试图在学校和周围环境之间建立一种联系，从而不会令使用者感到丝毫封闭感。玻璃被灵活的运用在各个不同方面：教室用的是净白片玻璃；大厅和咖啡厅用的是像在百货商店橱窗那样的大分格玻璃；花纹玻璃和不透明玻璃用于完全或部分阻隔视线干扰的地方。开放式的立面投射进来大量的光线。

这些光线还有其他的呈现方式，建筑本身看上去好像与地面没有联系，像一片叶子飘浮在那里。建筑室内铺设的木制地板和屋面的下表面是由热带爱洛克木材制成的。

Secondary School, Rozay-en-Brie

The secondary school of Rozay-en-Brie in the Département Seine-et-Marne resembles a small cité in its own right, located between a new residential neighbourhood and a supermarket. With its calm and easily legible exterior, the building is decisively set apart from its surroundings. With its “urban spaces”, it is clearly organised and has an obvious hierarchy, manifest, for instance, in the enclosed square as the central space and the open galleries for circulation. In terms of typology, a certain relation to the large farm estates of the region can be made out. The closed-off character of these estates, however, is superceded by great transparency here.

The main building is made up of a low square block, laid out like a cloister in its basic organisation. An axial walk traverses the school in a north-south direction and continues in the outdoors. To date, only the first building phase has been completed, so that the school is currently U-shaped in plan and the architects' conceptual idea cannot be experienced.

The totally glazed facades are a special feature, creating transparency even across the large interior courtyard. The architects managed to establish a relationship between the school and its surroundings that dissolves any feeling of being locked in. Glass is used in different ways: clear glass in the classrooms; large formats, as in storefronts, in the large hall and the cafeteria; serigraphed and opaque glass in rooms where visual connections are not, or only partially, desired. The open facades project great lightness.

This lightness is also expressed in other ways. The building itself seems unconnected to the ground, poised there “like a leaf”. Wooden floors and undersides of roofs are made of tropical Iroko wood.

▶ 学校位于小镇的边上，玻璃中庭由步廊环绕

The school is located on the edge of town and is present as a glass hall with surrounding walkways

在学校的中心庭院中，有一个容量更小的建筑，这是“学生中心”，里面设有独立于学校的学生专用咖啡厅和俱乐部。两个高耸的怪诞的体量像是为进行什么秘密活动提供隐蔽的空间，一个是完全封闭的，另外一个是朝北向开大窗，它们是学校的音乐和艺术教室。

报告厅略微倾斜的插入到学校的主体建筑中，平面呈梯形，在这个不可思议的“衣柜”里面容纳了600个学生的更衣柜。建筑虚实对比的设计主题反复再现，为学生营造了生动的学习环境。

In the centre of the school's square figure there is a further, smaller building. It is the student house. Here the students have there “own” cafeteria and club room independent of the school. Two towering foreign-looking volumes appear to harbour secretive activities-one is totally closed, the other has a large window facing north. These are the school's music and art rooms.

The auditorium is inserted into the large hall as a slightly slanted block, trapezoidal in plan–an enigmatic “suitcase” enclosed by 600 student lockers. The contrast of opening and closure, a recurring motif of this school building, can be vividly experienced by the students.

▶ 木材贴面的顶棚由纤细的圆柱支撑

The roof with a wooden underside is supported by slender round columns

▲ 两个高耸的体量里容纳了学校的音乐和艺术教室（上）

The two towering volumes conceal the music and art rooms

▲ 设有俱乐部的“学生中心”位于学校中心庭院里面（左下）

The “student house” with club rooms is located in the courtyard

► 报告厅插入到学校的主体建筑中，它的下部容纳了600个学生的更衣柜（右下）

The auditorium was inserted into the large hall and is enclosed by 600 lockers

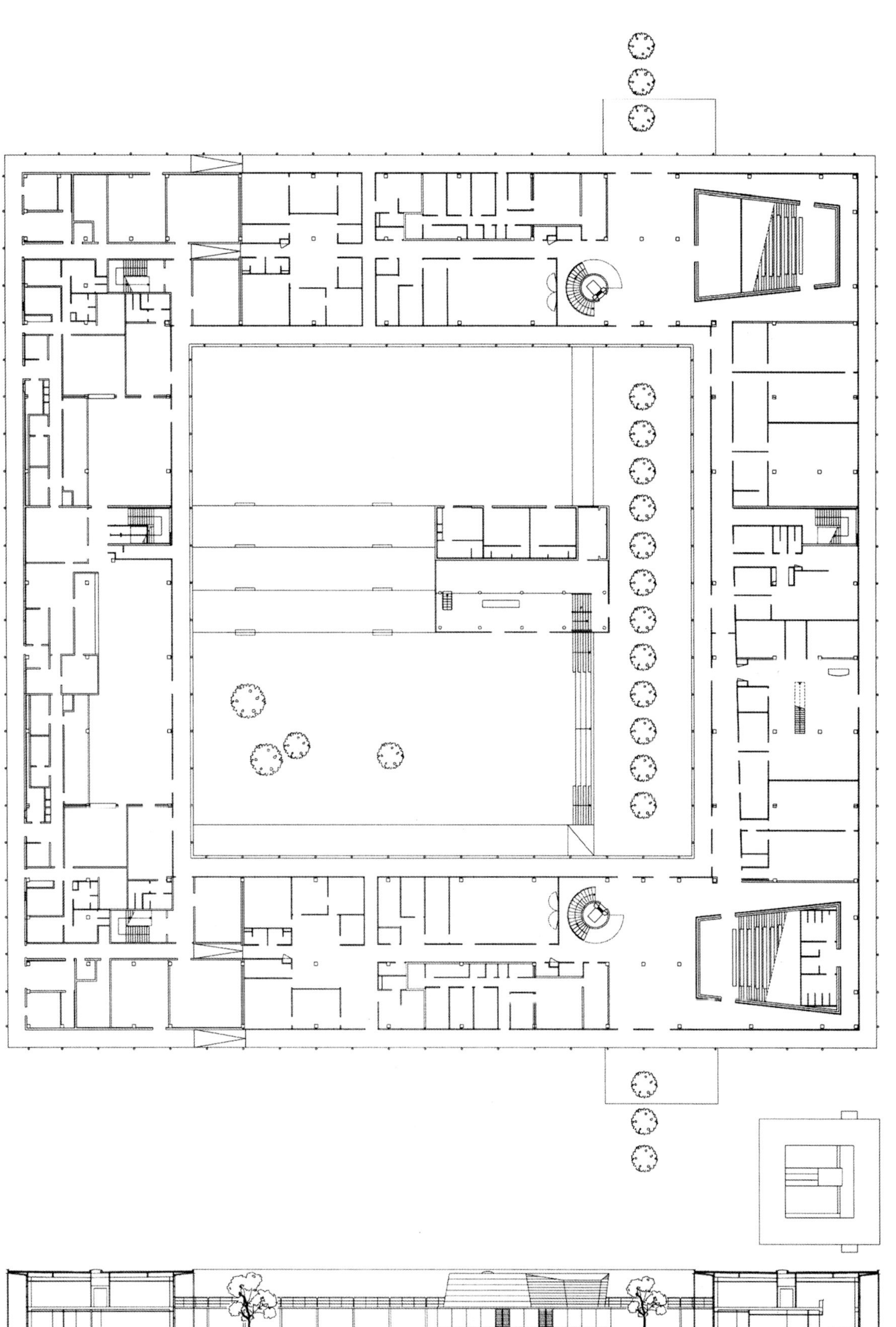

◄ 学校的平面图和剖面图，该建筑迄今为止只完成了一半

Plan and section of the school, to date only half-built

音乐厅和音乐学校，索恩河畔沙隆

Concert Hall and School of Music, Chalon-sur-Saône

索恩河畔沙隆是第戎南部的一个小镇。新的音乐中心对于整个地区有着特殊的意义，由于这是查隆——伯高尼交响乐团的发源地，新建筑将容纳主演奏厅和练习厅。这是在小镇的新区圣——考斯梅建设的第一批建筑之一，音乐厅还将在该地区发展过程中发挥重要的中心作用。

这个建筑综合体目前仍然有点孤立。两个主要功能——学校和音乐厅——具有很高的识别性。耐久性成为建筑师设计的主题。在一侧，教室空间看上去保守而封闭。舞蹈空间被安排在一个水面的后面，水面使过往的行人与舞蹈室保持一定视觉距离。在另外一侧，高技术特征的音乐厅看上去像一个“贝壳”建在空间紧凑的舞台和后台旁。穿过玻璃大厅，贝壳像是被展示在橱窗中。可改变和延伸的舞台正好可以适应为舞蹈表演伴奏的室内演奏乐的演奏需要。另设三个音乐工作室为录音和播音服务。

在这个建筑综合体内两个空间被建筑师用清楚的设计手法划分开来。插在两个体量之间的中央大厅，同时通向学校和音乐厅。两侧都要经桥状的楼梯方可到达此处。这条通道不仅起到交通的作用，而且还起到“动态光影”的视觉和轴线联系的作用。这个中介空间在这里起着重要的作用，它成功地在位于较远处的老城和正在实施中的新城之间实现了对话。

Chalon-sur-Saône is a small town south of Dijon. The new music centre is especially meaningful to the entire region, as it is the home of the Chalon-Bourgogne philharmonic orchestra, and the new building will constitute its primary performance and practice space. It was one of the first buildings to be completed in Saint-Cosme, a new part of town, and is to function as a prominent centre for the area in the process of development.

The resulting building complex is currently still somewhat isolated. The two functions—school and concert hall—can be identified immediately. The duality became the architects' project theme. On one side, the classroom volume seems reserved and largely closed. Dance spaces are set behind a water surface that establishes the necessary distance to passing pedestrians. On the other side, the highly technical concert hall appears like a "seashell" built next to the compact stage and backstage volume. Through the glazed lobby the shell can be viewed as if displayed in a shop-window. The variable and extendable stage is as suited for chamber concerts as it is for dance performances. Three sound studios are available for recording and broadcasting.

Two spaces of deliberate expressivity clearly separated from one another have thus been created. The central lobby, inserted between the two volumes, leads both to the school and the concert hall. It is reached from both sides across a bridge-like ramp of stairs. The path not only organises the circulation, but creates a visual, axial connection as a "ligne dynamique". This is important as it successfully creates a connection both to the old part of town, located a bit further away, and to parts of the new development yet to be implemented.

▲ 音乐厅内表面由白桦木装饰

The concert hall is concealed behind a birch wood shell

▲ 通透光亮的舞蹈室。水面使过往的行人与舞蹈室保持一定视觉距离

Glazed ballet studio. A water basin keeps passers-by at a distance

为了预防塞纳河的洪水，建筑没有设计地下室。建筑整体在一个高5m的方形底座上面，底座内设管理和技术用房。

正如当地市政所希望的，所选择的形式和材料使建筑成为显著的城市地标。它如一个奏响的“欢庆的器乐”带动了整个环境气氛。透过玻璃幕墙可以看到用白桦木质板装饰的音乐大厅的外饰面。浅色的木质板和入口大厅深色的镶木地板形成了强烈的对比。杰罗梅·布鲁奈特在这里设置了巨幅“曼陀林图画”。另一方面，学校的立面，可能会令人联想到一个符号系统，或是一个打点机压过的条带；孔洞呈自由状排列，但是与抽象的混凝土和金属铜相交错的条形结构相得益彰。

Because of the chance of the Saône flooding, the building does not have a basement. It is placed on a plinth five metres high containing the administration and technical services.

Forms and materials were chosen to explicitly mark the urban attractor so desired by the municipal client. A “festive instrument” with a far-reaching aura was to be created here. The concert hall volume, clad in bent birch wood panelling, is visible behind the glass facade. The shell’s light wood strongly contrasts the lobby’ s dark parquet flooring. Jérôme Brunet pictures “a huge mandolin in its case” . With the school’ s facade, on the other hand, one may think of a notational system, or the punched tape of a grinding organ; the holes appear to be placed at random, but fit into the abstract striped structure that results from the alteration of concrete and copper.

◀ 学校和音乐厅是建筑相互独立的两个部分

School and concert hall are two clearly separate building parts

▼ 音乐厅的侧立面

Side facade with concert hall

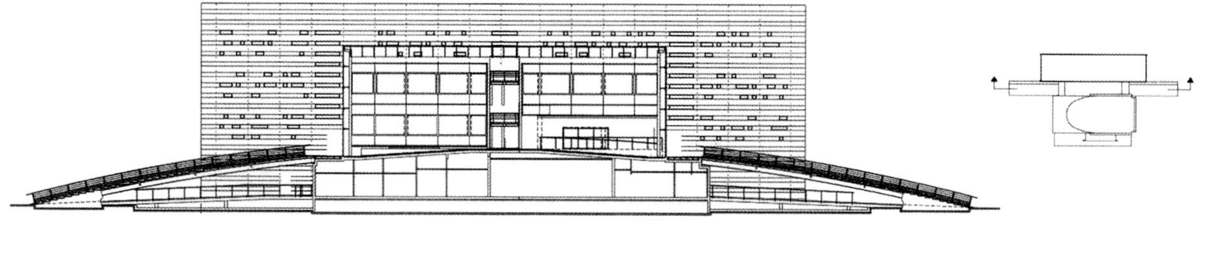

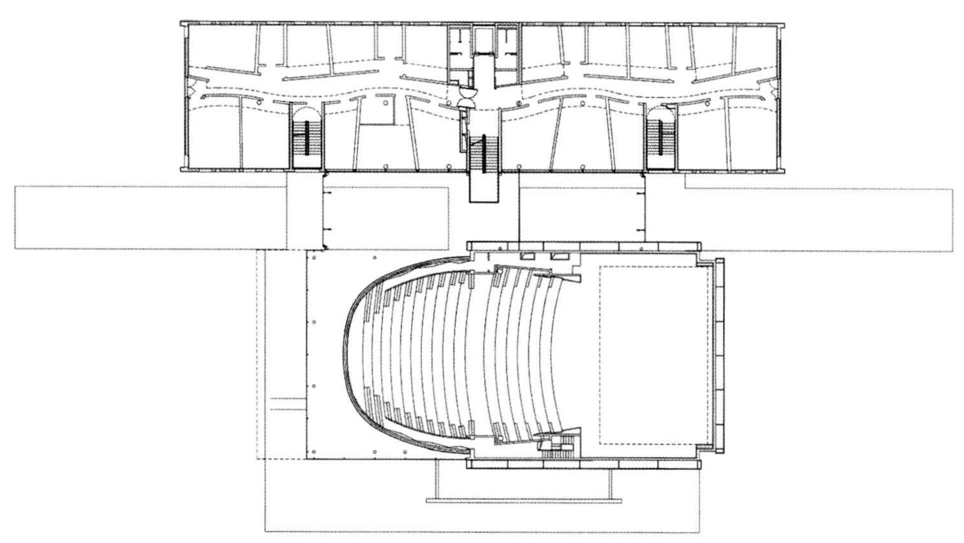

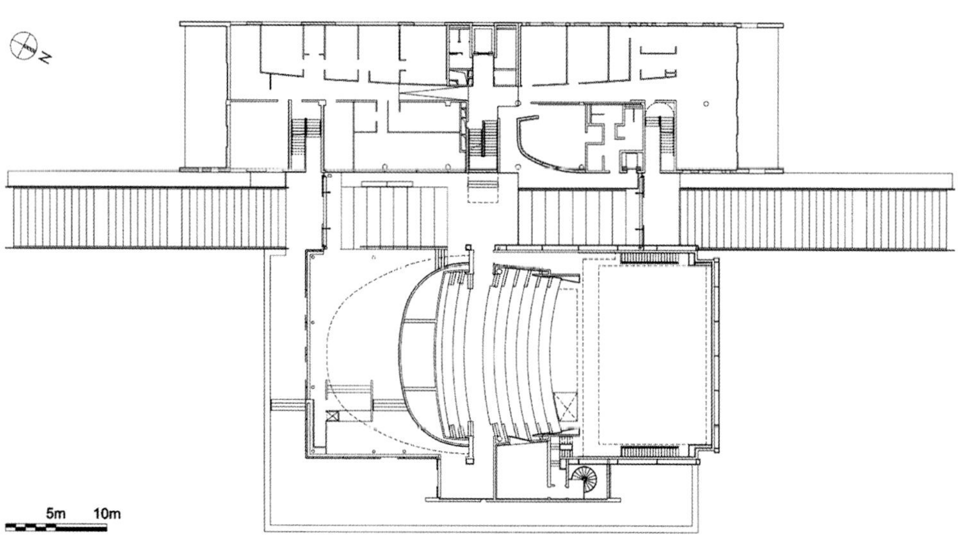

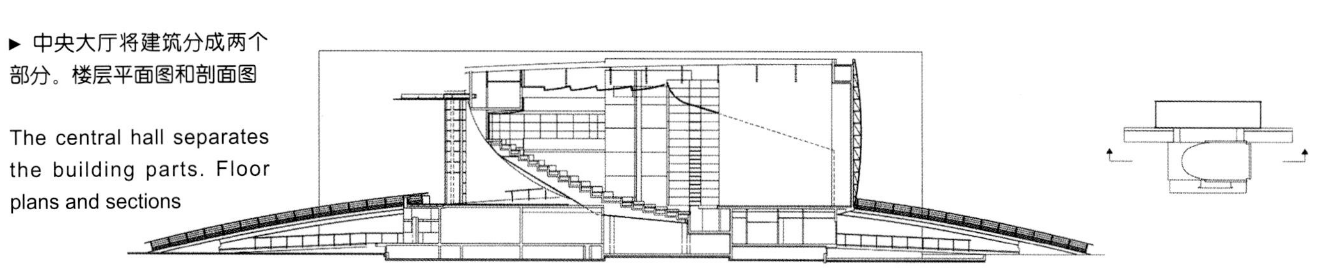

► 中央大厅将建筑分成两个部分。楼层平面图和剖面图

The central hall separates the building parts. Floor plans and sections

▲ 入口处丝毫不差地再现了经典建筑作品——巴塞罗那博览会德国馆

At the entrance, the Barcelona Pavilion as the building's model is unmistakable

等候空间位于左侧，相对应的银行交易大厅位于右侧。从长向立面射入的光线不仅照亮了大厅的前台，而且穿透到略微升高的后台工作区。在这巧妙的设计思想引导下，建筑师成功地完成了开敞办公区的一翼。此外，光线从屋面上的狭缝中透射进来，还有一个玻璃旋转楼梯通向二层办公区。

建筑的三个部分从功能上是分开的，但在形式上又紧密的结合在一起，建筑的第三个翼里设有饭店和通向北方的平台。从这里可以欣赏到周围坡地上美不胜收的茂密绿化和珍稀树木所构成的景致。在建筑主体内的

Inside, an extensive circulation zone with waiting areas is located to the left, while the banking hall lies to the right. This space is well-lit through the long facades, also at the back, where employees work on a slightly elevated level. With this astute concept, the architects succeeded in articulating the open office wing. Light is also admitted through a narrow slit in the roof and a glass spiral stair leading up to the office wing.

Functionally separate but integrated in terms of mass, the third building volume includes the restaurant and a terrace extending to the north. From here there is a beautiful view of the sloping terrain with its dense vegetation and the extraordinary trees of the neighbouring property. A bar within the main

▲ 银行的交易大厅在双向设置了落地窗，表现出娴静和高雅的气质

The banking hall is glazed on both sides and characterised by its sober elegance

酒吧与交易大厅和接待区用不透明的玻璃分隔开来。酒吧成为去办公空间下方报告厅的过渡区。各工会的委托办公室位于饭店的后面。法兰西银行这个建筑致力于从不同层次展现材料、透明度和令人兴奋的设计概念，以满足不同的用处。室内和室外流动空间的相互对话还在继续，同时建筑极大地享受了如公园般的环境景观。

building is delicately partitioned from the banking hall and the circulation zone by opaque glass. The bar creates the transition to an auditorium beneath the office wing. The mandatory offices of the different labour unions are located behind the restaurant. The Banque de France building owes its force to the materials, the transparency and the exciting concept of different levels with distinct uses. The dialogue continues in the relation of interior and exterior and takes advantage of the park-like surroundings.

▲ 两幅围绕在办公区周围的木制走道的实景

Two views of the wooden walkways that wrap around the office wing

◀ 员工休息区的平台

Terrace of the staff recreation area

▲ 从建筑后部的内庭院看西立面，一层是会议室

View from the west of the building's back section with conference halls on the ground floor

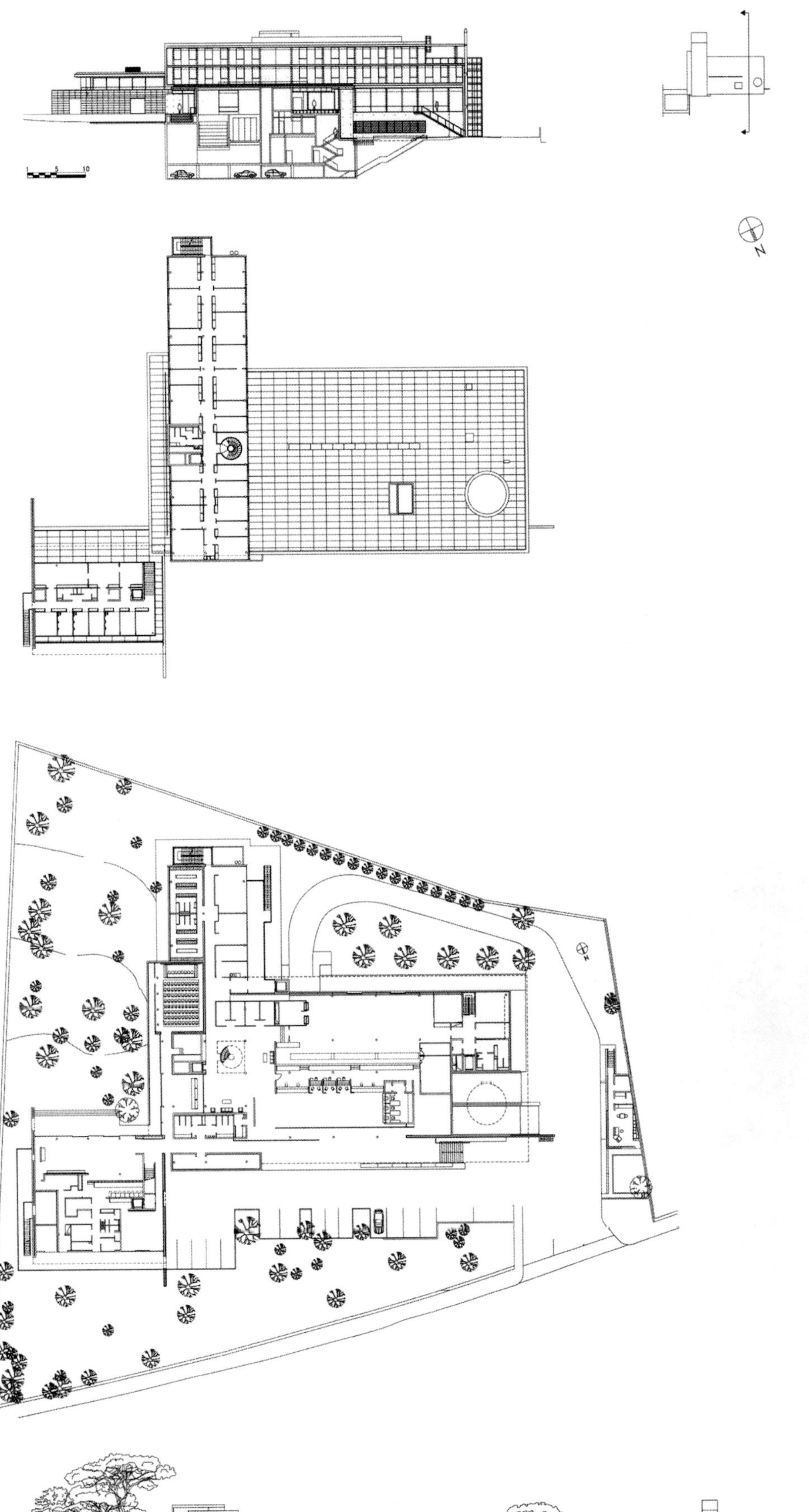

▶ 建筑剖面图，以及一层、二层的平面图和立面图

Section, ground and upper floor plans and elevation

在20世纪60年代，巴黎第六大学和巴黎第七大学形制严格的建筑建于圣路易斯的塞纳岛和植物园之间。该建筑群平面呈十字交叉的网格状，并由一个高起的体量控制全局。基地的北侧与阿拉伯世界研究中心相邻，该学院是由让·努韦尔在20世纪80年代建设的。在这之间还留有一个宽大的场地。在1992年，莱姆·库哈斯在此设计了一个玻璃图书馆，但是由于造价太高，最终未能实现。

布鲁奈特和索尼尔的设计任务包括在最短的时间和最少的预算内为学校营造灵活的研讨和办公空间。除此之外，还要提升该场地的城市空间品质。虽然，该设施只在未来有限的时间内——大学内其他建筑拆除石棉板工程完成以前——使用。

两个巨大的体量已经完成，都是43m长，四层楼高，每一层使用面积为750m^2，房间沿中间走廊两侧一字排开。除了卫生间和技术用房，研讨和办公的空间都以1.2m为模数。为了能够最大限度地实现空间组织上的灵活性，楼梯被布置在建筑的外侧，并与封闭的廊道相连。它们将新建筑与校园开放式的平台相联系，并在现有网格建筑的下方形成了一个连续的层面。最初，两部开敞楼梯计划设在两栋建筑之间的绿地上。但不幸的是，最后只能决定在建筑的端部各设一部楼梯。

Jussieu大学，巴黎

Jussieu University, Paris

In the sixties, the strictly rational buildings of Paris University VI and VII “Pierre et Marie Curie” were erected between the Seine island of St. Louis and the botanical gardens. The buildings, laid out on a lattice grid, create a distinct quarter dominated by a highrise. To the north, the area is bordered by the Institut du Monde Arabe, built by Jean Nouvel in the eighties. In between, a wide and deep space remained. In 1992, Rem Koolhaas proposed a glass library for this site, but because of high costs it had no chance of being built.

Brunet and Saunier' s task consisted in creating flexible seminar and office space for the university with minimal time and budget. In addition, the area's problematic urban qualities were to be upgraded. The facilities, however, were to be used only for a limited period of time, until asbestos removal in the surrounding existing university buildings is complete.

Two large blocks were completed, both 43 m long and with four stories of 750 m^2 usable floor space each, organised efficiently along a central corridor. Except for the sanitary and technical cores, the size of the seminar and office space is adaptable on an axis dimension of 1.2 m. In order to allow for maximum flexibility in spatial organisation, the stairs were located outside the buildings and are reached across enclosed bridges. They connect the new buildings to the open platform of the university campus, which forms a continuous level beneath the existing grid buildings. Originally, two open stairs were planned on either side of the green area between the two buildings. Unfortunately, it was later decided to place a stair at each short end instead.

▶ 玻璃幕墙赋予了建筑很强的表现力，只有走进才能看到隐藏在幕墙后面的立面

The glass elements lend the building original architectural force. Only upon getting closer does the facade behind become clearly visible

由于受时间和资金的限制，建筑在建造中完全使用了预制构件。考虑到防火安全的因素，结构由混凝土制成，而立面则是一个夹心的构造组合：标准的外窗和铝合金装饰条——朝外部的涂灰，朝向内院的涂白。两栋建筑最主要的特征就是竖向的瑞克莱特的工业化玻璃元素，片与片之间紧密相连，沿纵向附在每一层简单的立面外部。U形的断面沿最小的角度放置可以限制内外的视线交流。除了具有遮阳的功能之外，“玻璃立面”自身具有很重要的建筑艺术品质。从远处看，建筑外观呈现出抽象的状态，几乎是非物质的太空来物。

玻璃元素给予了不同的体量简单、容器式建筑以建筑艺术的原动力。当观察者受好奇驱使而走近建筑时，他或她会突然发现隐藏它后面的建筑的第二个层面，此时也就明白了建筑的真正用途。

Because of limited time and money, the buildings were assembled using prefabricated elements only. For fire-safety reasons, the structure was made of concrete, while the facades consist of sandwich elements with standardised windows and aluminium sheet, painted grey toward the outside, and white toward the courtyard. Crucial for the character of the two buildings are the Reglit industrial glass elements, mounted closely in front of the simple facades on every floor and across their entire length. The U-shaped sections are set at a slight angle, restricting views in and out. Apart from the function of shading, the design quality of the “glass facade” is the dominant feature. Seen from a distance, the building turns into an abstract, almost immaterial object of indeterminate use.

The glass elements endow the otherwise plain, container-like buildings with original architectural force. The observer approaches, driven by curiosity, until he or she suddenly makes out the second layer of the facade and can then understand what the buildings are about.

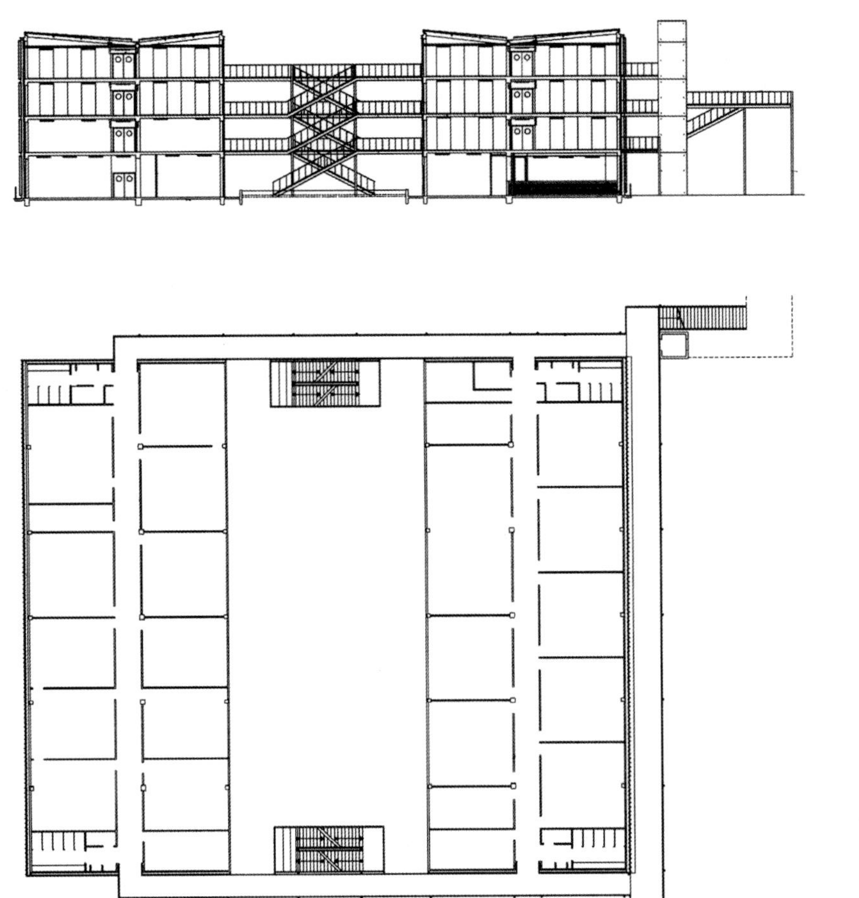

▶ 剖面图和楼层平面图

Section and floor plans

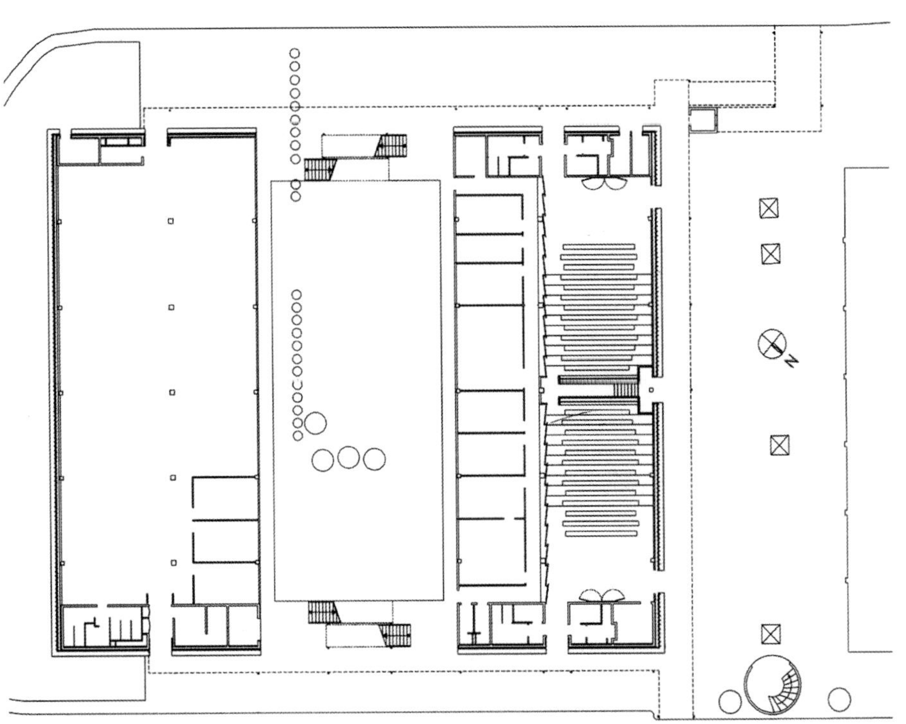

圣皮埃尔疗养院，帕拉尔弗莱洛

圣皮埃尔疗养院位于帕拉尔弗莱洛(Palavas-les-Flots)，这里是南蒙彼利埃地中海地区的知名小镇，是孩子们私密的疗养恢复佳地。这是一个拥有古老和传奇历史的地方。疗养院先是在沿海地带建立了一些小体量的建筑，进入20世纪疗养院得到了大笔的捐赠后又增建了一些三层的建筑，如宿舍、检查室、食堂和礼拜堂。在几次扩建之后，建筑最终总长度达到了240m。

由于原有疗养区在空间和技术上不再适应当代医疗的发展需要，因此规划建设一个新疗养区迫在眉睫。在招标委员会的组织下，有五位建筑师参加了概念性的设计竞标，并很快获得了最终的方案。在短短四年规划和建设中，新疗养区在1999年4月顺利完工。

新疗养区由一个沿街的两层黄色建筑，一个矮一些的连接栋和一个4层的深红色建筑构成。三幢建筑呈U形，南面朝向大海。混凝土楼板的出挑起到遮阳的作用。由于几个不同建筑体量外形各异，所以从大街一眼望过来并不会给初识者留下清楚深刻的印象。

建筑入口台阶表层材料是踩上去会吱吱作响的热带木板，它周围以矮树篱为屏。平面柱网呈正方形分隔，十字形柱的断面沿柱身向上逐渐变宽，底部以鞋状的金属基座锚固。静谧、舒适的前厅给来访者营造出宾至如归的温馨氛围。入口精心设计但并不奢华的木质装修地面向来这里疗养的孩子们展开温暖的怀抱，但却没有显露出疗养院的治疗气氛。

Saint-Pierre Institute, Palavas-les-Flots

Saint-Pierre Institute in Palavas-les-Flots, a small resort town on the Mediterranean south of Montpellier, is a private rehabilitation clinic for children. It has a long and eventful history. After modest beginnings in huts on the beach, in the twenties the donation of a large piece of property enabled the construction of a three-storey building with dormitories, examination rooms, a refectory and a chapel. After several extensions, the building finally reached a total length of 240 metres.

Planning a new clinic became necessary as the old complex no longer complied with today’ s requirements in terms of space and technology. With the support of a general contractor, an ideas competition was organised among five architecture practices and a decision was quickly made. After only four years of planning and construction, the clinic was inaugurated in April 1999.

The complex is made up of a two-storey yellow building volume along the road, a lower connecting zone, and a four-storey building volume, rendered in an oxblood colour, almost U-shaped in plan and open to the south toward the sea. Its concrete floor slabs project to shade the floors below. The first, slanted, view from the street leaves visitors with an unclear impression of several different building volumes.

Arriving visitors step between the buildings on to a creaking area planked with tropical Ipé wood and screened by a low pergola. The columns with cross sections, whose profiles broaden slightly upward, are set in a square grid and are anchored with steel “shoes”. This peaceful, almost cosy “anteroom” generates a feeling of being protected and well taken care of. The wooden arrival area deliberately makes no reference to a sanatorium and is – free of any daintiness – a space for children.

▶ 建筑傍水而立，包括恢复性医疗用房和三层的疗养病房

The building is located right on the beach and consists of a base with rehabilitation facilities and three floors of patient rooms

推开入口的推拉门，建筑真正的入口大厅是一个玻璃的中介空间，大厅西侧向外部绿色天井敞开，围合天井的上部是一个休息咖啡座，大厅引导前来就诊的病人向北到门诊室，向南到检查室，向西可以通过一个开敞式楼梯到达天井上部的咖啡座。大厅作为南部的联系枢纽，终点是海岸边的棕榈树庭院。

尽管如此，新疗养院建筑的真正魅力却完全集中在建筑上部的层次上。三层的病房区朝向大海呈U形展开，坐落在逐层出挑的平台上。病房区位于建筑的两侧短翼内，设有各种大小不同的病房。位于建筑中部退回的部分包括进餐区、游戏区以及工作人员用房。将孩子们与建筑周围的邻居隔离开，并最大可能地将水景引入房间，这对于布鲁奈特和索尼尔来说是很重要的，因此他们在每个房间前面都最大限度的营造出开放的空间，最下一层的病房外部的露台空间最为开阔。平台的上空沿着海岸设有廊架。从这里可以俯瞰两个种有棕榈树的内天井，它们同时是面向大海的疗养大厅的侧墙。

Behind a sliding door, the actual entrance lobby is a glazed intermediate zone which opens west on to a green patio. It connects to doctors' rooms for external patients to the north, and leads to examination rooms to the south. To the west, the patio is bordered by an elevated café terrace reached across an open stair. The lobby is the central connecting space to the south, culminating on the beach in a courtyard of palm trees.

The clinic' s real charme, however, is experienced only on the upper levels. Three floors of patient rooms are terraced and oriented to the sea in a U-shape. The shorter wings contain patients' rooms of different sizes, while the central, set-back section includes the dining and play areas as well as the staff rooms. It was important to Brunet and Saunier that the children be screened from their neighbours, and yet, being accomodated right on the water, that they have maximum open space directly in front of their rooms. The first floor of the patient wing is the most generous. The windows open on to a terrace with a pergola-covered walk along the sea. From here it is possible to look down into the two courtyards planted with palm trees that flank the sea-front therapy pool hall.

▸ 朝向大海的疗养病房俯瞰呈U形展开

The terraced patient wing opens to the sea almost U-shaped in plan

从海上远眺疗养院，修长的建筑体量很优雅的沿着海岸线展开。除了两处朝向内天井的开敞空间，建筑底层平面完全是由木制百叶构成的封闭外墙所围合。透过木制百叶墙的缝隙，室内空间的人只隐约显露出他的轮廓。两处朝向内天井的开敞空间可以被同样材质的木制百叶推拉门所封闭，因此疗养水池可以被完全封闭为内部空间，特别是在海滩使用的旺季。经过一段时间，棕榈树会长得比木制百叶墙还要高一些。

Seen from the sea, the long building is extraordinarily elegant. Except for two openings leading to the courtyards, the ground floor consists of a wall of wooden slats that forms a closed surface set in front of the actual building. People behind the wall are discerned as outlines only. The two entrance openings can be shut with sliding doors made of the same wooden slats, so that the pool can be entirely closed to the outside, especially during the busy beach season. The palm trees will grow higher than the wooden wall in the course of time.

▲ 每一个通向露台的病房都设有木制百叶推拉门，可使内部空间相对封闭

The individual patient rooms can be closed with sliding doors made of wooden slats

◀ 建筑前面的露台以及一层平面的棚架都是轮船甲板空间的再现

The terraces in front of the building and the pergola on the first floor are reminiscent of a ship' s deck

▶ 由一层治疗室围合的两个棕榈树庭院可以直接到达海边

The therapy rooms on the ground floor enclose two courtyards with palm trees with direct access to the sea

▸ 沿街与主体建筑相连的是一个建筑高度较低的入口大厅，以及与它相连的室外廊架和一幢辅助设施楼

A lower entrance hall with a pergola and a services wing toward the street are connected to the main building

▸ 设有廊架的内部庭院位于入口区的后侧

An interior courtyard is situated behind the entrance area with the pergola

▸ 大型的疗养水池完全朝向大海，外部饰以木制百叶墙以阻隔外部的视线

The large therapy pool is open to the sea. The wall of slats, however, allows only limited views from the outside in

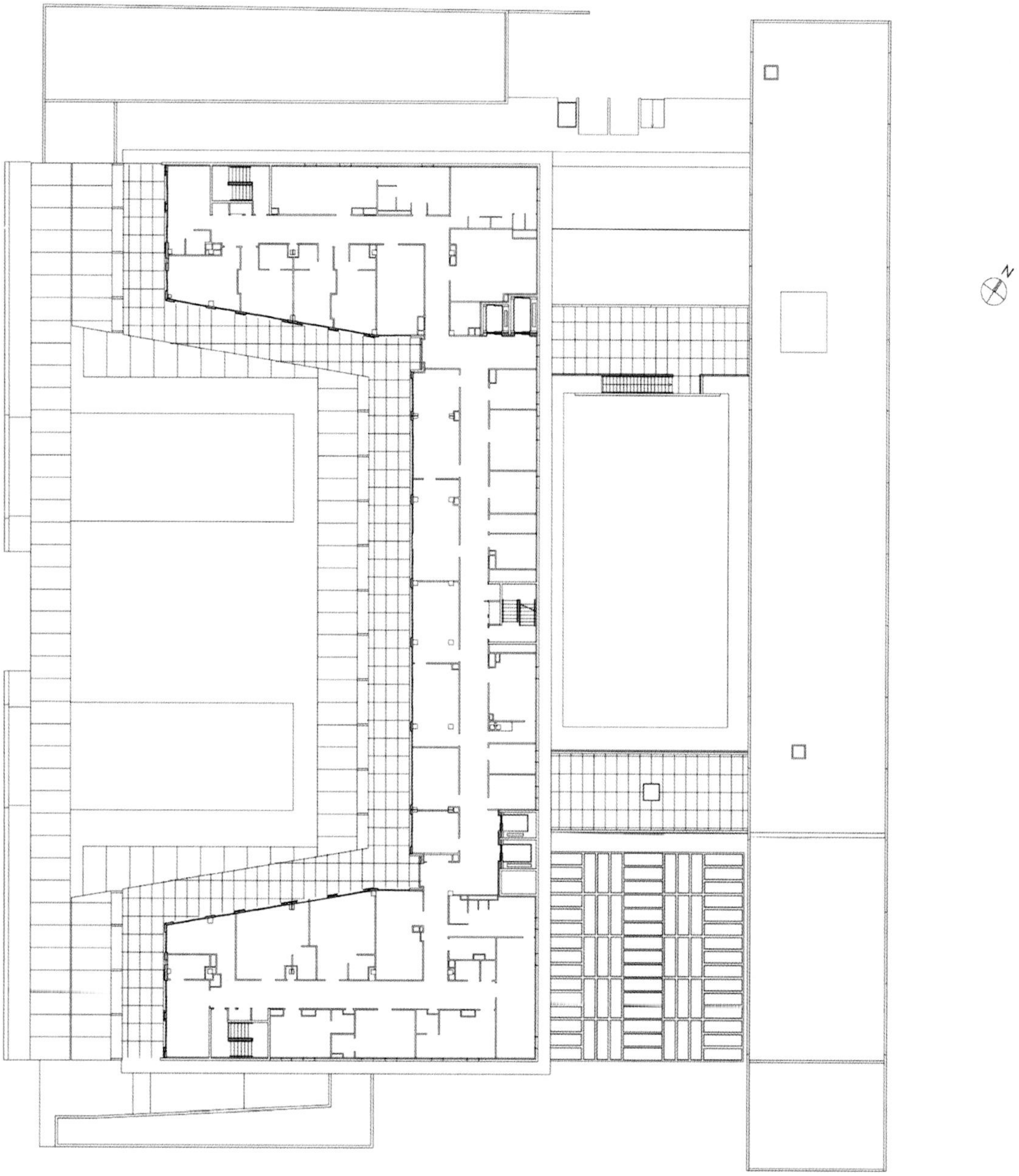

▶ 四层的平面图和剖面图

Third floor plan and section

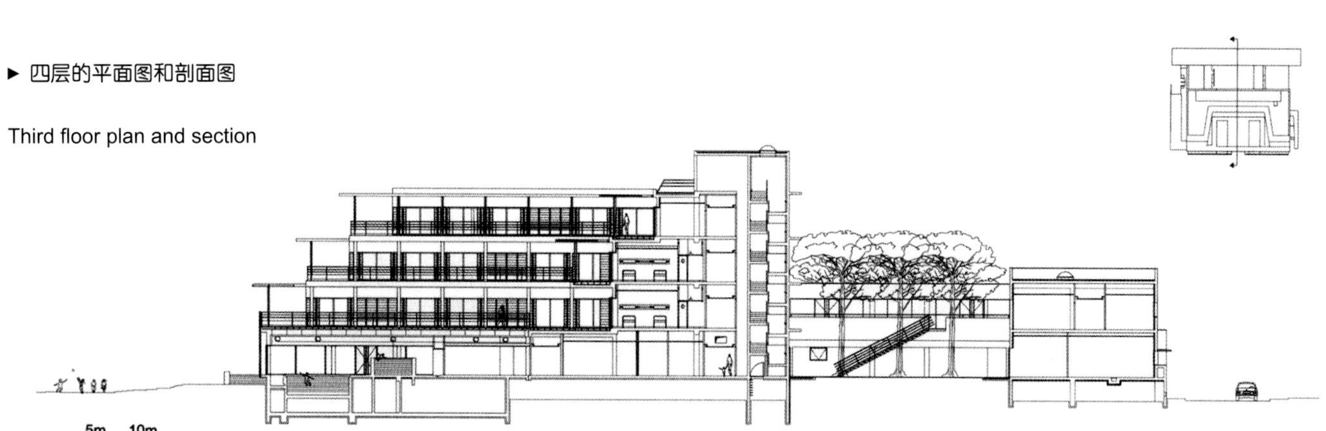

Neuve Tolbiac街公寓，巴黎

玻璃元素的运用成为该建筑最显著的特征。公寓位于Neuve Tolbiac街地区，与塞纳河成直角关系，面朝一个旧的冷库，这是一个现存极少的“与塞纳河不对称”式的老建筑。还有多米尼克·佩罗设计的法国国家图书馆的四个塔楼围绕在周围，这是一个重要的城市改造区，长度超过了3km，覆盖了130hm^{2}土地。根据1989年制定下来的规划设计指导要点，建筑朝向塞纳河的楼层都设有外挑的平台，因此，两个顶层的楼面都有巨大的屋顶平台。三个略微突出与建筑主体垂直相交的部分也同样设置了外部平台。

虽然该项目只是一个住宅公寓，但是环境建筑文脉和其紧临法国国家图书馆的重要地理位置使它运用奢华的处理手法显得顺理成章。尽管受到尺度和其他因素的约束，建筑师还是设法寻求到了为95户提供单元住宅的简单的空间解决方案，由于其良好的功能关系而使空间显得别具特色。虽然平面中出现了一些连续开敞的空间，但在多数情况下他们的平面设计还是不同寻常，横纵分明。很显然，该建筑设计受到了20世纪70年代简单实用的住宅设计理念的影响。这种理念在垂直交叉的平面设计，阳台以及大面积玻璃推拉门中得到了充分的体现。

外立面连续水平的带型阳台成为建筑标志性符号。贯穿整个建筑进深的入口大厅空间，尺度开阔而通畅。为了和竖向立面的韵律相一致，入口采用了减法空间并使它上方的跃层住宅略为后退。建筑外部水平阳台玻璃栏板雕刻着让·阿努伊

Residential Building Rue Neuve Tolbiac, Paris

The residential block uses glass as its most prominent feature. Positioned at right angles to the Seine on Rue Neuve Tolbiac, it faces an old cold-storage depot, one of the few original buildings remaining in “Seine Rive Gauche”. Surrounding the four towers of Dominique Perrault's Bibliotheque Nationale, this important urban redevelopent area is more than three kilometres long and includes 130 hectares of land. According to the design guidelines for the entire area approved in 1989, floors are terraced toward the Seine. Thus the two top floors have enormous roof terraces. The orthogonal building volume with its three slightly projecting extensions toward the block interior were also stipulated in the design brief.

Although the project is subsidised social housing, the architectural language and its location next to the Bibliothèque Nationale suggest luxury condominiums. Despite size limits and other norms, the architects managed to develop simple spatial solutions for the 95 flats, remarkable in particular because of good functional relationships. Nevertheless, despite some open spatial sequences in the flats, in many cases their layout is unaccustomed and cornered. Clearly, the architects designed a building influenced by the rational simplicity and legibility of housing blocks of the seventies. This is apparent above all in the orthogonal floor plans, the balconies and the large glass sliding doors.

The building is strongly characterised by the continuous horizontal bands of balconies. This clarity and generous scale is maintained in the lobbies that span the entire depth of the building. In order to comply with the vertical facade rhythm required by the design guidelines, the duplex flats above the entrances were slighty set back to create structuring incisions. The organisation of the building therefore becomes legible in the flight of

▶ 建筑大面积的玻璃外装修上雕刻的黑体字使建筑显得别有情趣。多米尼克·佩罗设计的法国国家图书馆在其后院

The largely glazed building is characterised by continuous bands of balconies with black type. Dominique Perrault's Bibliothèque nationale de France in the background

和皮埃尔·肖戴洛·德拉克所写的文章的片段。透过栏板建筑主体清晰可辨。周围一条新建道路以及穿过该社区的一条步行道就是以这两个作者的名字命名的。在玻璃栏板中采用的黑体字是为了纪念这个地区原来曾经有过的铁匠铺。

平面中沿横向展开通高的推拉门减少了来自外部窗户对室内直接的阳光辐射，而为室内空间带来了特殊的品质。尽管住宅单元平面面积较小，但是大面积的落地窗和特别的环境景观为在此居住的人们带来了一种感受，那就是这里也是可以充分享受生活的驿站。

the glass balcony parapets where fragments of texts by Jean Anouilh and Pierre Choderlos de Laclos were serigraphically applied. A surrounding new road and one of the pedestrian paths that traverse the block were named after the writers. The form of the black lettering on the balcony parapets wants to recall the blacksmiths' shops once located in the area.

A special quality in the flats is produced by the long rows of room-high sliding doors, made possible by lowering the radiators into the floor along the window fronts. In spite of their small size, therefore, thanks to the windows and extraordinary views, the flats can generate the feeling of being part of a lavish staging.

FOURNIL
jardin à la française
ROSES.
PREMIER TABLEAU
Laclos
Choderlos de L
liaisons

▶ 建筑平台朝向塞纳河。位于屋顶的单元可以享用两个大面积的屋顶平台

The building is terraced toward the Seine. The penthouse therefore has two very large roof terraces

▶ 从一个住宅单元远眺对面由弗朗西斯·索莱尔设计的建筑

The neighbouring building by Francis Soler seen from one of the apartments

▶ 开阔的入口大厅

A generous entrance lobby

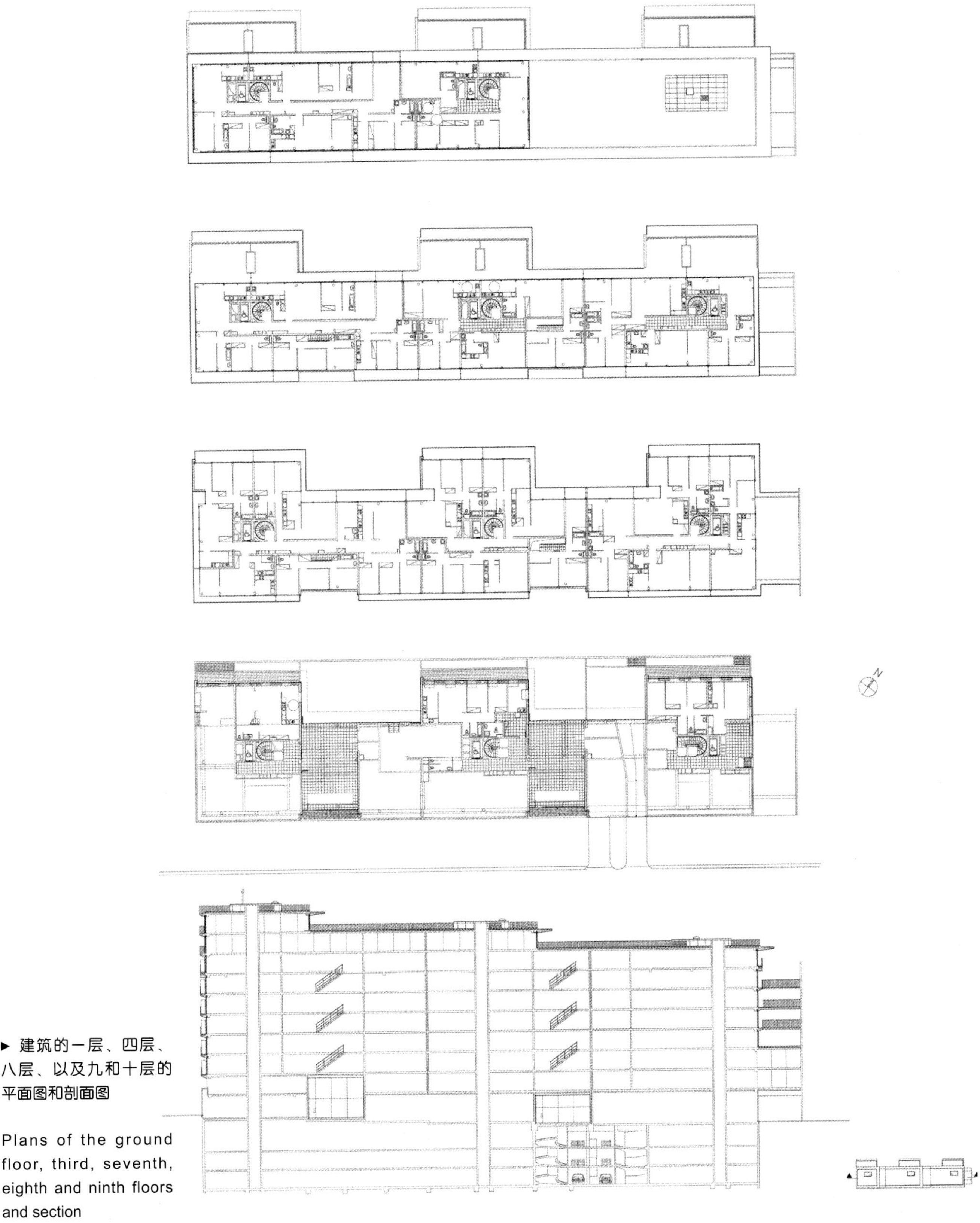

▶ 建筑的一层、四层、八层、以及九和十层的平面图和剖面图

Plans of the ground floor, third, seventh, eighth and ninth floors and section

弗兰德大街住宅，巴黎

弗兰德大街住宅位于拉维莱特湾(Bassin de la Villette)附近，在19世纪这里是劳动阶层生活的区域，该地区在20世纪60年代开始向北扩展。在此期间大量经历过世纪之交的老建筑被摧毁，取而代之的是高层住宅区，它部分延续了原有的沿街立面。在它南侧有几个大规模的街区，因为受到街道的阻碍而被夹在已有的建筑中间。由于在空间尺度上的差异，新建弗兰德大街住宅的建筑形式将表现得不同寻常。

新建住宅位于弗兰德大街的北端，紧邻一个穿越该地区的高架铁路线。杰罗梅·布鲁奈特和艾瑞克·索尼尔试图在这里使用他们从未尝试过的建筑语言。建筑的外形尺度和同质性使它与街道中其他大型建筑取得了平衡。其立面结构很明显是基于19世纪巴黎的住宅和办公建筑风格——这正是建筑师所说的“古典性”。

配有白色推拉百叶窗的落地玻璃窗不仅仅是立面的构成元素，而且组合起来就形成了具有韵律感的横向序列。沿着立面的竖向，横排的百叶窗仅在楼层处被楼板的外装饰带所打断。立面上落地百叶窗开合的随机性使立面形成了丰富的图示语言。基于此，建筑立面完全与环境脱离开来。它从根本上诠释了“巴黎人的住宅”的最初含义。

建筑最具争议的地方就是通高的落地窗。外立面所表现的连续楼板与内部的实际楼板高度并不相符。窗户上部采用的是深色反射玻璃，从而遮住了后面实际

Residential Buildings Rue de Flandre, Paris

Rue de Flandre, near the Bassin de la Villette in the working-class 19th arrondissement, was widened on its north side in the sixties. Numerous turn-of-the-century buildings were demolished and replaced by high housing blocks, that only partially follow the line of the street. On its south side, a few large blocks, set back from the street, were added between the existing buildings. Because of the resulting discrepancies in scale, Rue de Flandre has assumed unusual urban form.

The new building is located at the northern end of Rue de Flandre, next to a train viaduct that crosses it here. Jérôme Brunet and Eric Saunier attempted to frame the corner block in a language unusual for their practice. In its size and homogeneity, the building acts like a counterweight to the large buildings in the street. Its facade structure, “classical” as the architects would say, is clearly based on the Parisian residential and office buildings of the 19th century.

The large French windows with white sliding shutters are not only components of the facade here, but added to form long rows they are its constituent elements. Vertically, the rows of shutters are interrupted only by the narrow bands of the floor slabs. The free alteration of open and closed surfaces dissolves the facade and even transforms it into a graphic system. In this sense, the facade is clearly set apart from its surroundings. It is an entirely original interpretation of the “Parisian house”.

A debatable aspect of this building are the windows that cross floor height. The continuous floor bands outside do not correspond to the actual floor height on the inside. The upper glass zone of the windows, which masks the actual floor slab behind, is therefore

▲ 新建住宅位于一个重要的区域——弗兰德大街的北端，紧邻一个穿越该地区的高架铁路线

The corner building is located at a prominent site next to a railway bridge at the northern end of Rue de Flandre

的楼板，这在外部是觉察不到的。立面上出现的横向条带事实上是水平栏杆。由于栏杆的高度至少要1m，因此在上方落地窗的外侧加了一段玻璃栏板。真正的窗高仍然是2m，室内净高是2.6m。但是立面所表现的百叶窗开启的通高高度，内侧玻璃窗是达不到的。建筑在外观上遵循了传统建筑的比例，但实质却没有达到。

沿着铁路线共有两幢花园建筑，容纳了55个单元，沿阿戈讷大街的一部分尚未建设。通过一条狭窄的小路可以来到这两栋建筑，因为这条路贯穿了整个场地。开敞空间是由艺术家雅克里·多利亚克设计的。

made of dark mirrored glass. This is not, however, distinguishable as such from the outside. The horizontal band in the facade, then, is actually the parapet. As the parapet must be one metre high, it was necessary to add a pane of glass in front of the actual windows above. The actual window is still two metres high, and the clear height of the rooms measures 2.60 metres. But the generosity of the opening as celebrated in the facade is a fake. The building suggests historic proportions and is dishonest in doing so.

Including two garden buildings located along the railway tracks, a total of 55 flats were integrated into the block. One part of the ensemble along Rue de l'Argonne has not yet been built. The buildings are accessible along a footpath that extends across the full depth of the site. Open areas were designed by artist Jacqueline Dauriac.

▶ 新建住宅的立面语言成功地实现了与周围街区原有建筑的对话

The facade language successfully creates a relation to the neighbouring buildings in the side street

▶ 住宅的背立面与正立面大不相同，设有联系交通的平台和连廊，并插入了两栋低层建筑

Views of the back facade, very different from the front, with access galleries and terraces. Two low new buildings were also fit in here

▶ 窗户上部采用的是深色反射玻璃，从而遮住了后面实际的楼板（左）

The windows' upper glass zone, concealing the actual floor slab, is made of dark mirrored glass

▶ 在顶层的复式住宅单元（右）

One of the duplex flats under the roof

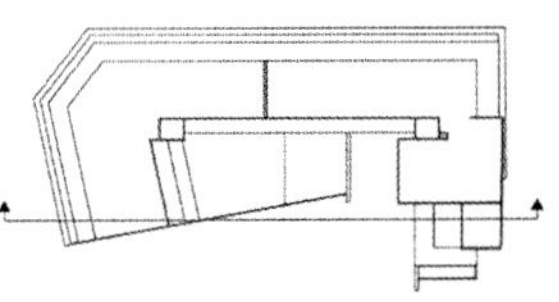

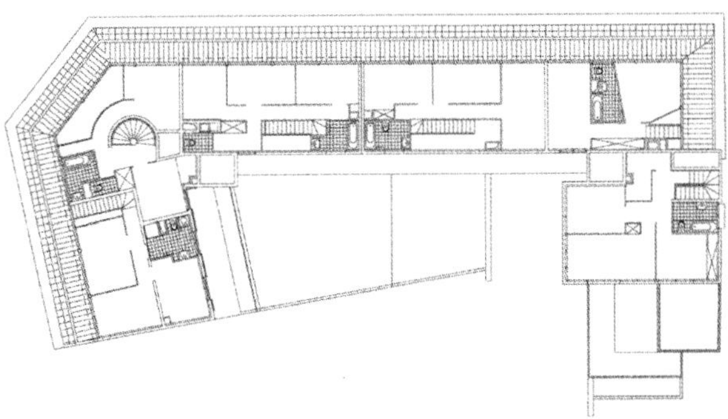

▶ 各层平面图和从庭院角度看的剖面图

Floor plans and section at courtyard

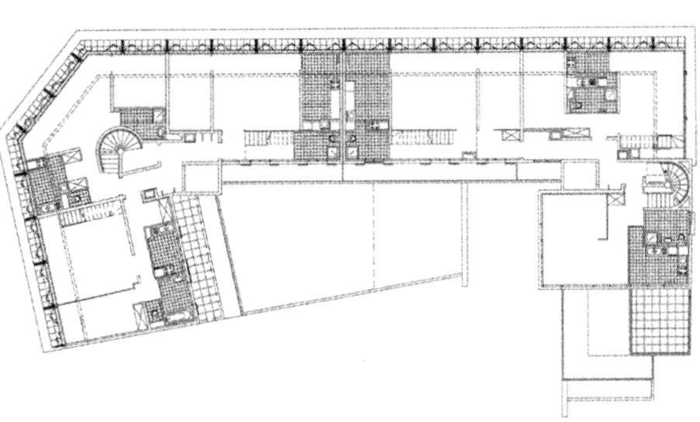

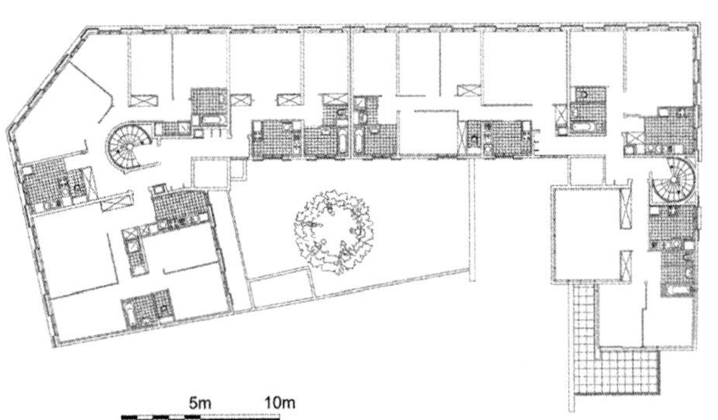

履历和作品展

Biographies and Selected Exhibitions

▲ 布鲁奈特和索尼尔在巴黎的事务所

Agence Brunet/Saunier, Paris

履历

杰罗梅·布鲁奈特1954年生于法国日索尔(Gisors)

1979年　毕业于法国国立高等艺术学校，获UPA7学士学位

艾瑞克·索尼尔

1952年　生于巴黎

1980年　毕业于国立高等艺术学校，获UPA6学士学位

1981年　成立布鲁奈特和索尼尔巴黎事务所

1982年　年轻建筑师获奖专辑

1991年　欧洲Italstat奖

班尼迪克特斯建筑奖——提名

1994年　法国巴黎博物馆研究所

1996年　圣日耳曼昂莱行政管理中心

AMO奖——提名

1998年　法国地方银行

2000年　圣皮埃尔学院，帕拉瓦莱弗洛

作品展

1982年　法国建筑师协会（IFA）——巴黎，Pan XII

1983年　Beaubourg中心——巴黎，第一建筑物

1987年　建筑公司——巴黎，Un Maire-un architecte

1990年　40位建筑师作品展

1991年　Beaubourg中心——巴黎，建造协会

1994年　维也纳，巴黎，伦敦，柏林——Glasbau Seele

1997年　军舰制造厂——巴黎，玻璃罩中的法国

1999年　魏玛的包豪斯学校，20世纪空间的变化

2001年　柏林，新空间

国家建筑师协会成员

国家建造协会质量咨询顾问

▲ 体育场馆，孔拉维尔市

SPORTS FACILITIES，Combs-la-ville

▲ 电信控制中心，弗莱尔

MAIN CENTRE OF TELECOMMUNICATION OPERATIONS，Flers

▲ 警察署，圣克莱尔市

POLICE DEPARTMENT，Hérouville Saint-Clair

建筑工程作品

Buildings and Projects

1983年

体育场馆
孔拉维尔市(Combs-la-Ville)，塞纳-马恩省(Seine-et-Marne)
竞赛——第一名
体育馆，舞蹈大厅，健身馆和附属建筑
业主：EPAMS
建筑技术：OTH Bâtiments
艺术家：Anton Solomouka
3 000 m^2——1986年竣工

1984年

国家图书管理中心
Saint-Lizier，阿列日省(Ariège)
邀请赛
位于13和14世纪主教礼堂内的书库，工作室和办公室
业主：文化部
工程管理：JB/ES & Edmond Lay
景观设计：Gilles Vexlard
建筑技术：OTH Bâtiments
7000 m^2

电信控制中心
弗莱尔(Flers)，奥恩省(Orne)
竞赛——第一名
研究室，办公室和住宅
业主：电信（Direction Regionale des Telecommunications de Caen）
建筑技术：OTH Bâtiments
2 000 m^2——1986年竣工

1985年

警察署
圣克莱尔市，卡尔瓦多斯省(Calvados)
业主：SGAP de Rennes
工程主持：Philippe Vasseur
模型：Eric Chicaez
工程技术经济：Patrick Guevel
800 m^2——1990年竣工

勃布朗市政大厅
勃布朗(Le Blanc)，安德尔省(Indre)
竞赛——第一名
地方、州管理办公室，议事厅、剧场
业主：勃布朗市
工程助理：Philippe Vasseur
特效：Daniel Darbois
建筑技术：OTH Bâtiments
3 000 m^2——未完成

▲ 倾诉塔，圣克莱尔市

GEINDRE TOWER，Hérouville Saint-clair

▲ 税务局，弗莱尔

TAX OFFICE，Flers

▲ 警察署，克雷泰伊

POLICE DEPARTMENT，Créteil

光通信和微波技术研究所
利摩日(Limoges)，上维埃纳省(Haute-Vienne)
竞标
业主：教育部
建筑技术：OTH Bâtiments

教育督察局
圣克莱尔市，卡尔瓦多斯省
竞赛——第一名
业主：教育部
工程助理：Eric Cugnart
建筑技术：Gay-Puig
4 500 m^2——1988年竣工

1986年

倾诉塔
或“欧洲最小的塔”
圣克莱尔市，卡尔瓦多斯省
业主：Francois Geindre
工程助理：Jean-Michel Reynier
65 m^2——1989年竣工

地方文化事务管理局
里昂(Lyon)，罗讷省(Rhône)
竞赛——第一名
设计内容：原有的军用医疗中心改为地方文化事务管理局（DRAC）
业主：文化管理局
建筑技术：OTH Bâtiments
5 800 m^2——未完成

PRET DE LA MANCHE 图书中心
圣洛(Saint-Lô)，芒什省(Manche)
竞赛——第一名
业主：文化部
艺术设计：Vincent Gontier
建筑技术：OTH Bâtiments
模型：Eric Chicaez
1 600 m^2——1988年竣工

1987年

VAL MAUBUEE学院
托尔西/马恩拉瓦莱(Torcy/Marne-la-Vallée)，瓦勒德马恩省(Val-de-Marne)
邀请参赛
业主：SAN
工程助理：Eric Cugnart
工程咨询：NEMO——A. Domingo 和 F. Scali
建筑技术：BETCI
5 500 m^2

税务局
弗莱尔(奥恩省)
竞赛——第一名
业主：财政部
工程助理：Isabelle Vasseur
艺术设计：Louis Derbré
结构工程师：Geciba——流体设计：Bethac
工程技术经济：Sery-Bertrand
模型：Eric Chicaez
1 200 m^2——1993年竣工

警察署
克雷泰伊(Créteil)，瓦勒德马恩省
竞赛——第一名
建筑内还包括警察署的其他服务部门
业主：内政部
工程管理：JB/ES & Delaage和Tsaropoulos
项目组成：Jean-Michel Reynier， Eric Dufour， Vitorio Pisu， Patricia Feraru
玻璃技术顾问：Marc Malinowsky
艺术设计：Francois Morellet
建筑技术：Serete
30 000 m^2——1992年竣工

LA MARE街住宅
La Mare街，巴黎第二十行政区，共有15个住宅、商店和停车单元
业主：OPHLM VP
工程助理：Philippe Vasseur（现场）
建筑技术：OTH Habitation
1 500 m^2——1989年竣工

▲ 地方市政厅，拉瓦勒

PREFECTURE OF MAYENNE，Laval

▲ 科学技术大学，略桑

UNIVERSITY INSTITUTE OF TECHNOLOGY，Lieusaint

▲ 坎帕拉法国大使馆，乌干达

FRENCH EMBASSY，Kampala (Uganda)

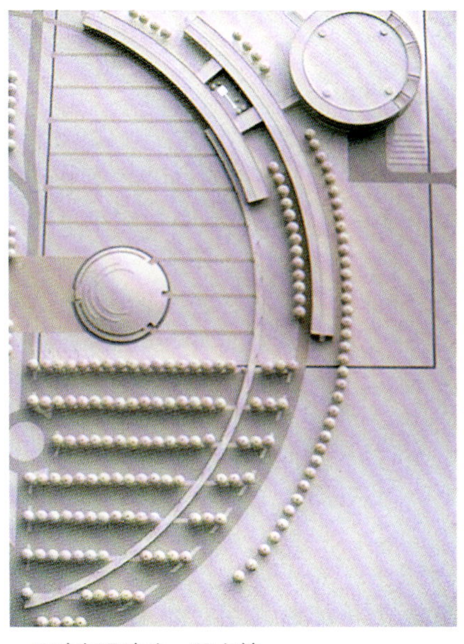

▲ 国家气象中心，图卢兹

NATIONAL METEOROLOGICAL CENTRE，Toulouse

1988年

地方市政厅
拉瓦勒(Laval)，马耶纳(Mayenne)
竞赛——第一名
建筑用途：用于满足为公众服务的市政办公所需要的空间
业主：内政部
项目组成：Jean-Michel Reynier，Eric Cugnart（竞赛），Jean-Paul Roynette（计划和现场）
艺术设计：Louis Derbré
建筑技术：Projetud
2 800 m^2——1991年竣工

科学技术大学
略桑(Lieusaint)，塞纳-马恩省
竞赛——第一名
建筑用途：市场研究部
业主：教育部
业主代表：SCARIF
项目组成：Laurent Jouannel，Edouard Grassin（竞赛），Jean-Michel Reynier(计划和现场)
建筑技术：OTH Bâtiments
3 000 m^2——1991年竣工

中央法院
卡昂，卡尔瓦多斯省
竞赛——第一名
设计内容：中央法院的改造和扩建
业主：司法部
项目组成：Hervé Bleton，Cécile Domin，Laurent Jouannel
工程咨询：Jean-Pierre Deschamps
建筑技术：OTH Bâtiments
6 600 m^2——未完成

职业训练中心
瓦朗谢讷(Valenciennes)，北部省(Nord)
竞标
建筑用途：教室、体育设施、停车场
业主：Valenciennes市
项目组成：Christian Basset，François Gillet
9 000 m^2

法国大使馆
坎帕拉 (Kampala)，乌干达(Uganda)
竞赛——第一名
业主：外交部
工程主持：Philippe Vasseur（竞赛和计划）
建筑技术：OTH Bâtiments
2 500 m^2——未完成

国家气象中心
图卢兹(Toulouse)，上加龙省(Haute-Garonne)
竞标
建筑用途：办公、计算机中心和会议中心
业主：工业装备部、房屋建设部、国土资源部、运输部；国家气象局
项目组成：Colette Brice，Patricia Ferreru，Jean Michel Reynier，Philippe Vasseur
建筑技术：OTH Bâtiments
15 000 m^2

PLACE DE SAVIES住宅
Place de Savies，巴黎第二十行政区
建筑用途：80个单元的住宅、商店和停车场
业主：OPHLM VP
工程主持：Philippe Vasseur
建筑技术：OTH Habitatin
8 000 m^2——1990年竣工

巴黎工业建筑
Porte d'Aubervilliers，巴黎第十九行政区
设计／建筑竞赛－第一名
设计内容：机械工业部技术服务建筑的改建及内部结构改造
业主：巴黎市，Direction de l' Architecture
项目组成：Christian Basset，Francois Gillet（竞赛），Patricia Ferraru（策划）
承包方：SPIE建筑
效果图：Alberto Bali
8 000m^2——1991年竣工

▲ 国际会议中心，巴黎

INTERNATIONAL CONFERENCE CENTRE, Paris

▲ 国际学校，圣日耳曼昂莱

INTERNAT LONAL SCHOOL, Saint-Germain-en-Laye

▲ NECKER医院，巴黎

NECKER HOSPITAL， Paris

▲ 卢佛尔宫研究实验室，巴黎

LOUVRE RESEARCH LABORATORIES, Paris

1989年

法国地方电信总部
卡昂，卡尔瓦多斯省
竞标
建筑用途：办公、会议和训练中心，停车场
业主：法国电信
工程助理：Agence “Avant-Travaux”“智慧建筑”工程咨询：Sophie Brindel-Beth
结构工程师：Becebat——流体设计：CET
工程技术经济：Sery-Bertrand
7 000 m^2

国际会议中心
Quai Branly，巴黎第七行政区
竞标
业主：外交部
工程主持：Jean-Michel Reynier
项目组成：Eric Babin，Frederique Guillouet，Clare Lasbrey，Jean-Francois Renaud，Isabelle Vasseur
特效：Daniel Darbois
结构工程师：Aroca-Marc Malinowsky
效果图：Vincent Lafont，Louis Paillard
70 000 m^2

“LA PLACETTE”小学
Place de l’ Oratoire，尼姆，加尔省(Gard)
竞赛——第一名
建筑用途：位于一个古镇，一个日护中心和一所小学的综合体
业主：尼姆市
工程主持：Jean-Michel Reynier
项目组成：Eric Babin，Jean-François Renaud，Isabelle Vasseur
结构工程师：Geciba——流体设计：Bethac
金属－材质：Arcora
工程技术经济：Sery-Bertrand
效果图：Louis Paillard
模型：Eric Chicaez
5 000 m^2——1991年竣工

圣日耳曼昂莱国际学校
Rue du Fer à Cheval，圣日耳曼昂莱，伊夫林省(Yvelines)
竞赛——第一名
业主：教育部
业主代表：DDE des Yvelines
工程主持（计划）：Jean-Michel Reynier（餐馆、剧院、图书馆）
Christelle Julien（中学、高中、日护部）
Paveo Kornecki（小学）
Piotr Glegola（停车场）
项目组成：Eric Babin，Jean-François Renaud, Isabelle Vasseur, Colette Brice
景观设计：Florence Mercier
建筑技术：OTH Bâtiments
改造面积：12 000 m^2
新增面积：10 000 m^2
1992～1995年竣工

NECHER医院
Rue de Sevres，巴黎第七区
竞赛——第一名
业主：国家救援中心
工程主持：Jean-Michel Reynier
建筑技术：Bethac
工程技术经济：Ripeau
前期：ODM
1 800 m^2——1992年竣工

卢佛尔宫研究试验室
卢佛尔宫，巴黎第一区
竞赛——第一名
建筑用途：位于Jardins du Carrousel地下的试验室
业主：卢佛尔宫公共管理机构
工程主持：François Guidon（策划），Isabelle Vasseur（场地设计），Valérie Faugeras（家具）
玻璃结构工程师：Focqué
建筑技术：OTH Bâtiments
效果图：François Robin, Vincent Lafont
5 500 m^2——1993～1995年竣工

▲ 音乐博物馆，巴黎

MUSEUM OF MUSIC，Paris

▲ 小块住宅区，阿涅尔

ILOT DE LA STAIION HOUSING，Asnières

▲ 文化中心，阿尔福维尔

CULTURAL COMPLEX，Alfortville

国家高等电信学校
Rue Vergniaud——rue Barrault，巴黎第13区竞赛——第一名
建筑用途：两栋建筑要满足管理和会议大厅，以及培训用房
业主：电信部
工程主持：Jean-Paul Roynette和Philippe Lair（竞赛），Vincent Marchand（策划和场地）
项目组成：Daniel Guetta，Hervé Levaseux
建筑技术：OTH Bâtiments
工程技术经济：Ripeau
5 200 m^2——1991年竣工

1990年

社区医疗中心
Verdun大街，克雷泰伊，瓦勒德马恩省
竞标
建筑用途：医疗技术区的附属建筑
业主：医疗中心
工程主持：Vincent Marchand
项目组成：Jean-Michel Reynier，Xavier Lagurgue，Clare Lasbrey，Thomas von Amelunxen
建筑技术：OTH Bâtiments
效果图：Louis Paillard
13 000 m^2

LAFARGE-COPPEE 社会服务中心
Rue des Belles Feuilles，巴黎第16区
邀标
建筑用途：办公
业主：Kaufman & Broad
工程主持：Jean-Michel Reynier
项目组成：Sophie Nicolas，Thomas von Amelunxen, François Guidon
效果图：Didier Ghislain
模型：Eric Chicaez
6 500 m^2

法国电力（EDF）培训中心
楠泰尔(Nanterre)，上塞纳省(Haute-de-Seine)
竞标
建筑用途：培训用房和工作室
甲方：EDF
项目组成：Catherine Dormoy，Elisabeth Lemercier
建筑技术：OTH Bâtiments
7 000 m^2

音乐博物馆
Cité de la Musique，巴黎第19行政区
竞标
博物馆概念设计
业主：Parc de la 市公园管理机构
工程主持：Catherine Dormoy
灯光：L' Observatoire1–Georges Berne
特种玻璃设计：Guillaume Saalburg
建筑技术：Bethac
工程技术经济：Ripeau
效果图：Vincent Lafont
4 500 m^2

小块住宅区项目
阿涅尔(Asnières)，上塞纳省
邀标
建筑用途：160个住宅单元和50个住宅单体
业主：SMCI Ile-de-France
工程管理：JB/ES与Dominique Perrault合作
项目组成：Catherine Dormoy，Lionel Loris，Line Sattler
效果图：Vincent Lafont
24 000 m^2

文化中心
阿尔福维尔市(Alfortville)，瓦勒德马恩省
竞标
建筑用途：1 500座剧院，300人的声乐表演练习厅，多媒体图书馆，展示空间
业主：Ville d' Alfortville
工程管理：JB/ES与Jean Michel Wilmotte合作

▲ 税务局，尼姆 TAX OFFICE，Nîmes

▲ E.N.S.M.A，普瓦捷

E. N. S. M. A.，Poitiers

▲ 教育局，鲁昂

EDUCATIONAL OFFICE，Rouen

项目组成：Catherine Dormoy，Frigo，Jean-Michel Reynier
特效：Richard Peduzi
建筑技术：Serete
效果：Louis Pailard
22 000 m^2

巴黎社区体育馆
Rue des Rigoles，巴黎第20区
建筑用途：将一个波尔塔式的大厅改建成为邻里服务的体育场馆
业主：巴黎市，建筑指导委员会，SLA20
工程主持：Isabelle Vasseur
助理：Danièle Forte
建筑技术：Arcora，Bethac
工程技术经济：Ripeau
1500 m^2——1992年竣工

1991年

社会科学大学
斯特拉斯堡(Strasbourg)，莱茵省(Bas-Rhin)
邀标
建筑用途：容纳哲学系，语言学系，信息系，艺术系和音乐系
业主：教育部，青少年体育协会，Strasbourg学术委员会
项目组成：Catherine Dormoy，David Long，Danièle Forte
声学设计：Lasa
建筑技术：OTH Est
效果图：Vincent Lafont
8 000 m^2

税务局
尼姆，加尔省
设计\建造竞赛
建筑用途：办公，接待
业主：法兰西内阁
工程管理：JB/ES与Nicolas Crégut合作
工程主持：Jean-Michel Reynier
建筑技术：OTCE
工程承包：BEC建筑公司
插图：Vincent Lafont
模型：Eric Chicaez
6 200 m^2

ROUEN教育局
Rue de Fontenelle，鲁昂(Rouen)，滨海塞纳省(Seine-Maritime)
竞标
建筑用途：校长办公室的改扩建
业主：教育部，青少年体育协会，Rouen学术委员会
项目组成：Igor Demidoff，Stanislas Melun
建筑技术：Geciba，Bethac
工程技术经济：Ripeau
效果图：Vincent Lafont
模型：Eric Chicaez
8 600 m^2

技术学院
略桑，塞纳-马恩省
竞赛——第一名
建筑用途：工业修理及维护系
业主：教育部
业主代表：SCARIF
工程主持：Jean-Michel Reynier
艺术设计：Anna Pricoupenko
建筑技术：OTH Bâtiments
3 000 m^2——1992年竣工

航空机械高等学校（E.N.S.M.A）
普瓦捷(Poitiers)，维埃纳省(Vienne)
竞标
建筑用途：办公室，工作室，教室
业主：教育部，青少年体育协会，Poitiers学术委员会
工程主持：Jean-Michel Reynier，David Long
建筑技术：Serete
模型：Eric Chicaez
21 000 m^2

▲ 外国公民接待中心，博比尼

FOREIGN CITIZENS' RECEPTION CENTRE，Bobigny

▲ 管理中心，圣日耳曼昂莱

ADMINISTRATIVE CENTRE，Saint-Germain-en-Laye

▲ 高中，尼姆

HIGH SCHOOL，Nîmes

▲ 国立音乐舞蹈学校，索恩河畔沙隆

ECOLE NATIONALE OF MUSIC AND DANCE，Chalon-sur-saône

外国公民接待中心
博比尼(Bobigny)，塞纳-圣但尼省(Seine-Saint-Denis)
竞标
建筑用途：办公和接待区
业主：内政部，塞纳-圣但尼省
工程主持：Vincent Marchand
助理：Jean-Michel Reynier
屋面设计：Marc Malinowsky
建筑技术：OTH Bâtiments
模型：Eric Chicaez
8500 m^2

行政管理中心
Rue Leon Desoyer，圣日耳曼昂莱，伊夫林省
竞赛——第一名
建筑用途：办公和公共服务设施，停车场
业主：圣日耳曼昂莱市
竞赛项目组成：Igor Demidoff
工程主持：Jean-Michel Reynier（策划），Isabelle Vasseur（场地设计）
玻璃结构工程师：Marc Malinowsky
家具：Eric Pouget
建筑技术：OTH Bâtiments
效果图：Vincent Lafont
7000 m^2——1994年竣工

RENNS山谷中央指挥港，车库，生产车间
Ligne J.F. Kennedy/La Poterie，
雷恩(Rennes)，伊勒-维莱纳省(Ille-et-Vilaine)
竞赛——第一名
业主：SEMTCAR
工程管理：JB/ES与“Avant-Travaux”合作
建筑技术：Matra
10 000 m^2——1999年竣工

1992年

尼姆高中
Bd Salvador Allende，尼姆，加尔省
竞标
建筑用途：容纳2 500名学生的高中
业主：Languedoc-Roussillon地区
工程管理：JB/ES与Nicolas Crégut合作
工程主持：Jean-Michel Reynier
项目组成：Nathalie Brilman，Alain Debords
结构工程师：René Vial——流体设计：Logibat
工程技术经济：Algoé
效果图：Thierry Lacoste
模型：Eric Chicaez
10 000 m^2

钢铁厂——制模车间
Technocentre Renault，Guyancourt，伊夫林省
竞标
建筑用途：炼铸生铁
业主：RENAULT SA
项目组成：Malcolm Nouvel，Elsa Cortesse-Vincent
工程技术经济：Delta
模型：David Topani
10 000 m^2

国立音乐舞蹈学院
Saint Cosme区，索恩河畔沙隆，索恩-卢瓦尔省(Saône-et-Loire)
竞赛——第一名
建筑用途：礼堂，教室，舞蹈演播室，电子视听室
业主：索恩河畔沙隆市
工程主持：George Yiontis
项目组成：Bach Nguyen，Victor Fuentes
特效：Michel Rioualec
声学设计：Lasa
家具：Valérie Faugeras
景观：Méristème
立面设计顾问：Jean-Louis Besnard
结构工程师：Terrell Rooke
流体设计：Bethac
工程技术经济：Delta
制图：Bernard Baissait 和 Cie
6 000 m^2——1995年完成

▲ 技术中心试验室，Guyancourt

TECHNOCENTRE LABORATORIES，Guyancourt

▲ 展厅，Pavillon de L'Arsenal，巴黎

EXHIBITION Pavillon de l'Arsenal，Paris

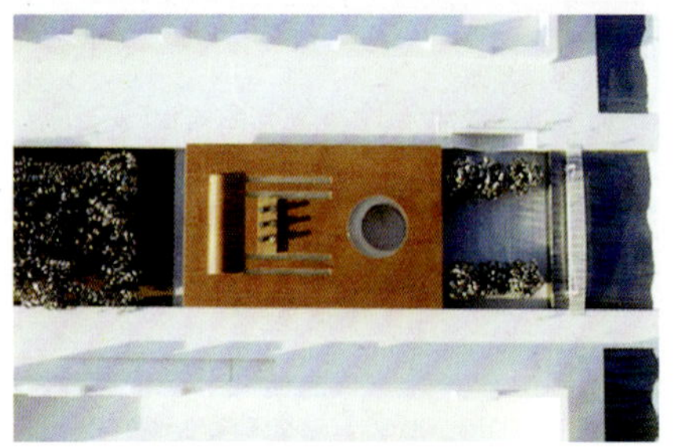
▲ 技术中心餐厅, Guyancourt

TECHNOCENTRE RESTAURANT, Guyancourt

▲ 高中，罗宰昂布里

HIGH SCHOOL，Rozay-en-Brie

▲ 路易斯 波登基地，埃尔内

LOUIS DERBRE FOUNDATION，Ernée

法国电信－OCTAL
蒙彼利埃，埃罗省(Hérault)
建筑用途：办公
业主：法国电信
工程管理：JB/ES与Nicolas Crégut合作
项目组成：George Yiontis，Bach Nguyen
结构工程师：René Vial——流体设计：Logibat
工程技术经济：Algoé
效果图：Vincent Lafont
模型：David Topani
3 500 m^2

技术中心实验室
Guyancourt，伊夫林省
竞赛——第一名
建筑用途：试验室，工业大厅，办公室
业主：Renault SA
竞标成员：George Yiontis
工程主持：Igor Demidoff，Vincent Marchand，Jean-Michel Reynier
项目组成：Lionel Renouf，Geneviève Shater
模型：David Topani
30 000 m^2——1998年竣工

技术中心餐馆
Guyancourt，伊夫省
竞标
业主：Renault SA
竞标成员：Jean-Michel Reynier
厨房设计顾问：SA Huron
流体设计：Inex
工程技术经济：DELTA
模型：Jean-Michel Françoise
3 000 m^2

“住宅百年”展览
巴黎第四区，Pavillon de l'Arsenal
“每个楼层的水和煤气”
特效，展览标识
业主：巴黎市民
助理：Philippe Vasseur
展览代表：Jacques Lucan
800 m^2——展览日期：1992.9~1993.1

LA TOUR DES DAMES高中
罗宰昂布里，塞纳-马恩省
建筑设计竞赛：第一名
业主：Region Ile-de-France
业主代表：DDE de Seine-et-Marne
工程主持：Vincent Marchand
项目组成：Malcolm Nouvel(竞标)
Pawel Kornecki(策划和场地设计)
艺术：Françoise Seigneur
建筑技术：OTH Bâtiments
工程承包：Dezelus建设
Chambon
效果图：Vincent Lafont
一期：6 500 m^2——1993年竣工

LOUIS DERBRE基地
埃尔(Ernée)，马雅讷省(Mayenne)
业主：Conseil General de la Mayenne-Ernee市
建筑用途：生产车间，铸造厂，LOUIS DERBRE 雕像的展示空间
助理：Isabelle Vasseur
模型：Eric Chicaez
1 000 m^2——1993年竣工

1993年

LIEUSAINT技术学院(3、4系)
默伦-塞纳尔(Melun-Sénart)，塞纳-马恩省
竞赛——第一名
业主：教育部
业主代表：SCARIF
工程主持：Jean-Michel Reynier
助理：Vincent Mégrot
建筑技术：OTH Bâtiments
7 500 m^2——1993年竣工

CACHAREL总部
Le Colisée，尼姆，加尔省
设计内容：Kisho Kurokawa建筑内部办公室和展示空间的重新设计
业主：Société Cacharel
工程主持：Christophe Kuntz
助理：Valérie Faugeras(家俱)

▲ CACHAREL总部，尼姆

CACHAREL HEADQUARTERS，Nîmes

▲ 法兰西银行， 蒙彼利埃

BANQUE DE FRANCE，Montpellier

▲ 巴黎卢佛尔宫学院，巴黎

LOUVRE SCHOOL，Paris

▲ 法国文化中心，路易港

FRENCH CULTURAL CENTRE， Port-Louis

▲ 巴黎LE SOU医疗总部，巴黎

LE SOU MEDICAL HEADQUARTERS，Paris

室内灯光、声音、标识：LM Communiquer
雕塑：Gatimalau
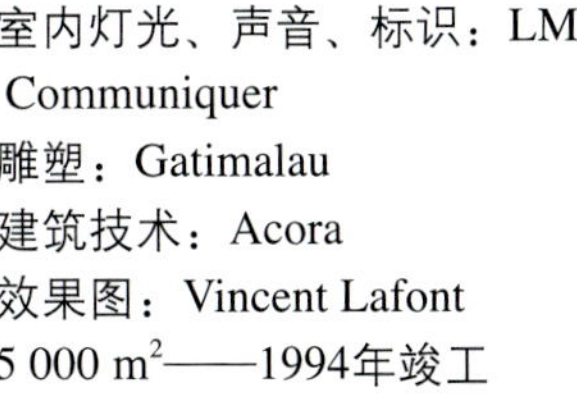
建筑技术：Acora
效果图：Vincent Lafont
5 000 m^2——1994年竣工

法兰西地方银行总部
Lodève大街，蒙彼利埃，埃罗省
竞赛——第一名
建筑用途：法兰西地方银行总部
业主：法兰西银行，不动产指导委员会
工程主持：Christophe Kuntz，George Yiontis，Bach Nguyen
助理：Jacques Lévy-Bencheton
立面设计顾问：Jean-Louis Besnard
混凝土设计顾问：Jean-Pierre Aury
家具：Valérie Faugeras
建筑技术：Serete建设
效果图：Anne和Denis Cleary
雕塑：Gatimalau
6 500 m^2——1996年竣工

巴黎卢佛尔宫学院
巴黎第一区，Aile de Flore-卢佛尔宫
竞标
建筑用途：地面层：报告厅，办公室，教室和图书馆，多层地下室
业主；卢佛尔宫公共建设部
助理：Philippe Vasseur
建筑技术：OTH Bâtiments
效果图：Vincent Lafont
4 200 m^2

中央法院
格拉斯(Grasse)，滨海阿尔卑斯省(Alpes-Maritimes)
竞标
建筑用途：特等审判庭，高等审判庭，经济审判庭，劳资调解委员会
业主：司法部
项目运作管理：DDE des Alpes-Maritimes
项目组成：Myriam Dao，Jean-Michel Reynier
结构工程师，YRM——流体设计：Inex
工程技术经济：Delta
效果图：Didier Ghislain，Vincent Lafont
20 000 m^2

法国文化中心
路易港(Port-Louis)，毛里求斯(Mauritius)，印度洋
竞标
建筑用途；展览空间，餐馆，工作室，会议大厅，剧场
业主：合作部
工程管理：JB/ES与Lampotang和Siew合作
助理：George Yiontis
结构工程师：OTH国际
效果图：Vincent Lafont
2 600 m^2

第戎教育中心
大学校园，第戎(Dijon)，科多尔省(Côte-d'Or)
竞标
建筑用途：办公室和公共设施
业主：教育部和第戎学术委员会
助理：Igor Demidoff
景观设计：Kathryn Gustafson
结构工程师：YRM——流体设计：Inex / Ingespie
工程技术经济：Delta
12 000 m^2

LE SOU医疗总部
巴黎第九区： Bellefond大街
设计内容；将一个原有的仓库改造成为办公空间
业主：SOU医疗保险机构
工程主持：Vincent Marchand
助理：Pawel Kornecki
结构工程师：Geciba——流体设计：Bethac
工程技术经济：Philippe Talbot
1 000 m^2——1994年竣工

▲ 急救中心，Le Blanc-Mesnil

FIRST AID CENTRE，Le Blanc-Mesnil

▲ 法国贸易和工业会所，巴黎

FRENCH CHAMBER OF COMMERCE AND INDUSTRY，Paris

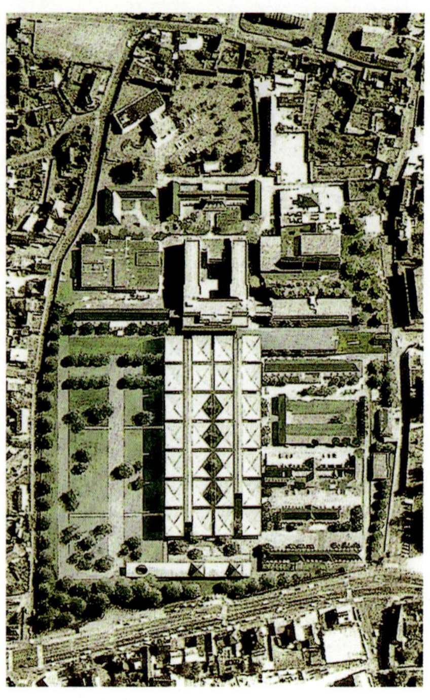

▲ 大学医疗中心，卡昂

UNIVERSITY HOSRITAL CENTRE，Caen

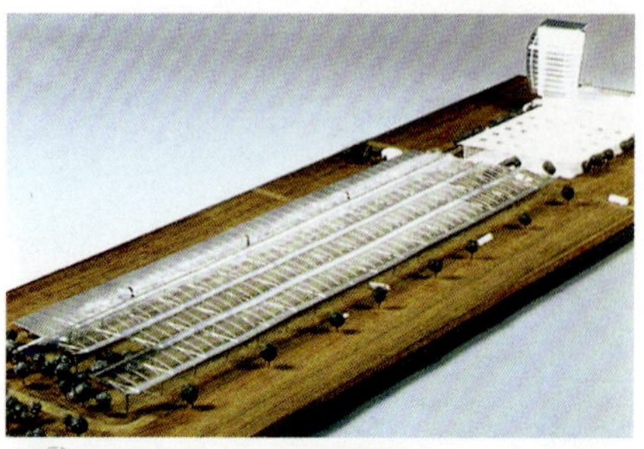

▲ GLASBAU SEELE工厂和办公室，盖斯特霍务

GLASBAU SEELE Factory and offices，Gersthofen

▲ 德意志联邦政府，柏林

FEDERAL CHANCELERY，Berlin

医科大学
图尔，安德尔和卢瓦尔省(Indre-et-Loire)
业主：图尔教区长
项目组成：Vincent Marchand, Igor Demidoff, Vincent Megrot
建筑技术：IMC
4 500 m^2

1994年

急救中心
Le Blanc-Mesnil，塞纳-圣但尼省
竞赛——第一名
业主：地方公安部门
工程主持：Jean-Michel Reynier
结构工程师；Geciba——流体设计：Bethac
工程技术经济：Delta
效果图：Vincent Lafont
模型：David Topani
5 000 m^2——1998年竣工

法国贸易和工业组织会所
巴黎第14区，圣雅依大道
竞标
建筑用途：办公室，会议中心，餐馆
业主：SCI Saint-Jacques ACFCI
业主代表：SCIC-AMO
土地整治工作：S.A.D.M
工程主持：Jean-Michel Reynier，Igor Demidoff
建筑技术：Sfica
效果图：François Robain
模型：Jean-Michel Françoise
8 500 m^2

大学医疗中心
卡昂，卡尔瓦多斯省
竞赛——第一名
业主：大学医疗中心
工程主持：Toshio Sekiguchi(竞标)，Jean-Michel Reynier(策划)
项目组成：Vincent Megrot，Federico Masotto，Pierre-Emmanuel Droste，Victor Fuentes，Adelaide Borniche
建筑技术：Serete 建设
效果图：Vincent Lafont
14 000 m^2———期1997年竣工

GLASBAU SEELE 工厂和办公室
盖斯特霍务(Gersthofen)，德国
邀标
业主：Glasbau Seele
工程主持：Matthias Bauer
建筑师：Vincent Marchand
项目组成：Jean-Michel Reynier，Igor Demidoff，Jacques Levy-Bencheton, Michel Plaxine
结构工程师：Alto(Marc Malinowsky)
流体力学：Kuehn Bauer+合作者(慕尼黑)
灯光设计：Christian Bartenbach(慕尼黑)
声学设计：Lasa
效果图：Anne和Denis Cleary
30 000 m^2

德意志联邦政府
柏林，德国
邀标
业主：德意志联邦政府
项目组成：Igor Demidoff，Jean-Michel Reynier，George Yiontis，Matthias Bauer，Pierre Emmanuel Droste，Vincent Megrot，Victor Fuentes
景观设计：Philippe Niez, Alexandra Schmidt
结构工程师：RFR
立面设计：ARCORA
多媒体设计：Pierre Miquel–Environ
效果图：Vincent Lafont
110 000 m^2

HENRI SELLIER高中
利夫利加尔冈(Livry-Gargan)，塞纳-圣但尼省
建筑设计竞赛
业主：Region Ile-de-France
工程主持：Vincent Marchand
建筑技术：Technip Seri 建设
工程承包：Dezellus建设
效果图：Vincent Lafont
8 000 m^2

▲ 住宅建筑，巴黎

RESIDENTIAL BUILDING，Paris

▲ 博物馆，圣·皮埃尔和密克隆岛

MUSEUM，Saint-Pierre et Miquelon

▲ 坎帕拉法国使馆二期，乌干达

FRENCH EMBASSY (2nd)，Kampala (Uganda)

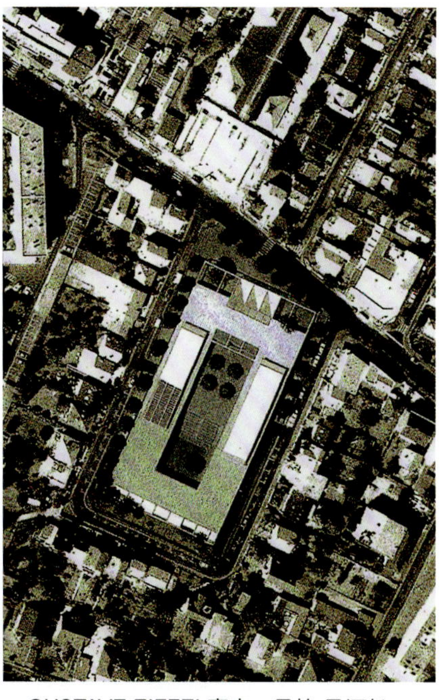

▲ GUSTAVE EIFFEL高中，吕埃-马迈松

GUSTAVE EIFFEL HIGH SCHOOL，Rueil-Malmaison

NEUVE TOLBIAC大街住宅
巴黎第13区Neuve Tolbiac 大街
竞赛——第一名
设计内容：95个住宅单元
业主：La Sabliere
工程主持：Vincent Marchand(竞赛)，Jean-Michel Reynier(策划)，Isabelle Vasseur(场地设计)
项目组成：Isabelle Vasseur，Bach Nguyen，George Yiontis，Federico Masotto，Valérie Faugeras
艺术设计：Anna Pricoupenko
结构工程师：Geciba——流体设计：Bethac
工程技术经济：Delta
模型：David Topani
10000 m^2——1997年竣工

车辆检测中心
Guyancourt，伊夫林省
竞赛
业主：Renault SA
项目组成：Igor Demidoff，George Yiontis，Lionel Renouf
建筑技术：Sogelerg
工程技术经济：Delta
模型：David Topani
基底面积：12 500 m^2

1995年

圣·皮埃尔和密克隆岛博物馆
圣·皮埃尔(Saint-Pierre)，法国海外领土
竞赛
建筑用途：博物馆和档案馆
业主：合作部
业主代表：SODEPAR
业主成员：SCET DOM-TOM
助理：Igor Demidoff
博物馆学：Jean-Francois Bodin
建筑技术：OTH国际
效果图：Vincent Lafont
2 500 m^2

GUSTAVE EIFFEL 高中
吕埃-马迈松(Rueil-Malmaison)，上塞纳省
竞赛
业主：法国地方议会 SEM92
工程主持：Vincent Marchand
项目组成：Jean-Michel Reynier，Vincent Megrot, Jacques Lévy-Bencheton
建筑技术：Technip Seri 建设
效果图：Vincent Lafont
模型：David Topani
12 000 m^2

法国使馆（二期）
坎帕拉，乌干达
竞赛
业主：外交部
建筑技术：Sincoba
效果图：Jacques Aillaud
2 000 m^2

TERRAGE-BOUTRON住宅
巴黎第10区
竞赛
业主：巴黎公共建设办公室
工程主持：Jean-Michel Reynier
工程技术经济：Philippe Talbot
效果图：Vincent Lafont
2 400 m^2

巴黎住宅建筑
Flandre大街，巴黎第19区
竞赛——第一名
业主：巴黎公共建设办公室
工程主持：Isabelle Vasseur
助理：Jean-Michel Reynier
建筑技术：OTH住宅
效果图：Didier Ghislain
5 200 m^2——1999年竣工

▲ 圣·皮埃尔学院，帕拉瓦莱弗洛

SAINT-PIERRE INSTITUTE，Palavas-les-Flots

◀ 住宅建筑，巴黎

RESIDENTIAL BUILDINGS，Paris

VALEO
利摩日，上维埃纳省
设计意图：Valeo地区的现代化改造
业主：Valeo(摩擦材料)
工程主持：Vincent Marchand
结构工程师：Terrell Rooke
流体力学：Bethac
工程技术经济：Philippe Talbot
4 000 m^2

ANDRE MALRAUX大学
阿涅尔，上塞纳省
竞赛
业主：Hauts-de-Seine议会
项目组成：Vincent Marchand，Vincent Megrot
7 600 m^2

圣·皮埃尔学院
帕拉瓦莱弗洛，埃罗省
邀标——第一名
建筑用途：儿童医院，康复治疗
业主：Montpellieraine儿童事务委员会(O.M.E.M)
业主代表：SCIC 发展
工程主持：Jean-Michel Reynier(策划)，Christophe Kuntz(场地设计)
项目组成：Jacques Levy-Bencheton，Federico Masotto，Alberto Medem
标识设计：Sabine Rosant S.R.P.C
景观设计：Christine和Michel Pena
建筑技术：Jacobs-Serete
效果图：Jacques Ailaud
13 000 m^2——1999年竣工

CACHAREL，巴黎
Etienne Marcel大街，巴黎第1区
建筑用途：收藏品展示厅
业主：Sté Cacharel
工程主持：Christophe Kuntz
助理：Valérie Faugeras
300 m^2——1995年竣工

法国电力办公室(EDF)
Quai Dedion Bouton，皮托(Puteauxa)，上塞纳省
竞赛
建筑用途：EDF——办公室
业主：EDF
工程主持：Igor Demidoff
建筑技术：Betom
11 000 m^2

1996年

PROSPER CHUBERT医疗中心
瓦讷(Vannes)，莫尔比昂省(Morbihan)
竞赛
设计内容：医疗中心改扩建
业主：Prosper Chubert医疗中心
工程主持：Vincent Marchand
项目组成：Eduardo Arroyo，Alberto Medem
景观设计：Christine，Michel Péna
建筑技术：Sogelerg
效果图：Jacques Aillaud，Vincent Lafont
40 000 m^2

LA PAILLADE维修中心
蒙彼利埃，埃罗省
蒙彼利埃郊区的第一条有轨电车线
竞赛，机修车间
业主：SMTUM
助理：Christine，Michel Péna
建筑技术：GEC Ingénierie
效果图：Jacques Aillaud
8 600 m^2

D'HENNEMONT城堡
圣日耳曼昂莱国际高中，伊夫林省
设计内容：改建整修一座19世纪古堡：报告厅，接待厅，住宿
业主：教育部
业主代表：DDE des Yvelines
工程主持：Philippe Vasseur(策划)，Isabelle Vasseur(场地设计)
结构工程师：Geciba——流体设计：Bethac

▲ D'HENNEMONT城堡，圣日耳曼昂莱

CHATEAU D'HENNEMONT，
Saint-Germain-en-Laye

▲ 大学医疗中心，兰斯

UNIVER SITY HOSPITAL CENTRE，Reims

▲ 地方事务多功能图书馆，香槟区沙隆

MUNICIPAL LIBRARY FOR REGIONAL AFFAIRS，Châlons-en-Champagne

工程技术经济：AEI
6 000 m^2——2000年1月竣工

RHONE税务局
里昂，罗讷省
竞赛
设计内容：税务局计算机中心改建
业主：财政部
工程主持：Vincent Marchand
建筑技术：Bethac
工程技术经济：Philippe Talbot
5 000 m^2

1997年

LYCEE FRANCAIS
法兰克福，德国
竞赛
业主：外交部
项目组成：Wolfgang Dievernich，Jean-Michel Reynier，Jacques Lévy-Bencheton，Vincent Megrot
建筑技术：Sogelerg，Lahmeyer GmbH国际
效果图：Jérôme Sigwalt
8 000 m^2

高环境标准(H.Q.E)高中建筑计划
科德里(Caudry)，北部省
建筑设计竞赛
业主：北部加来海市
项目管理：JB/ES & Graph建筑
工程主持：Vincent Marchand
景观设计：Bocage景观
建筑环境技术：Alter Ego
环境设计合作方：Pierre Diaz-Pedregal
承包方：Rabot-Duthilleul
效果图：Karine Herman和Jérôme Sigwalt/ Philippe Harden
12 000 m^2

地方事务多功能图书馆
香槟区沙隆(Châlons-en-Champagne)，马恩省(Marne)
竞赛
业主：香槟区沙隆市
项目组成：Jean-Michel Reynier，Jacques Lévy-Bencheton，Vincent Megrot
建筑技术：Technip Seri 建设
效果图：Vincent Lafont，Philippe Harden
6 800 m^2

大学医疗中心
兰斯(Reims)，马恩省
竞赛
建筑用途：IGGOMEP遗传学学院，妇科－产科和产期医药护理
业主：大学医疗中心
项目组成：Vincent Marchand，Gerold Zimmerli，Jean-Michel Reynier
建筑技术：Serete 建设
效果图：Philippe Harden
模型：David Topani
11 000 m^2

木工房
阿讷西(Annecy)，上萨瓦省(Haute-Savoie)
建筑用途：生产车间和展厅
业主：d'Agencement 技术
工程主持：Christophe Kuntz
建筑技术：Bethac
效果图：Philippe Harden
1 600 m^2——1998年竣工

预制安装建筑
Jussieu大学校园，巴黎第5区
建筑设计竞赛－第一名
建筑用途：为不断消除大学建筑中石棉制品而临时建设的校舍
业主：巴黎六大 Pierre et Marie Curie
业主顾问：G3A

▲ QUERTIER BOUFFLERS-FERRARE酒店，枫丹白露

QUARTIER BOUFFLERS-HOTEL DE FERRARE，Fontainebleau

▲ 建筑学院，蒙彼利埃

SCHOOL OF ARCHITECTURE，Montpellier

▲ BAUER体育馆，圣旺

BAUER STADIUM，Saint-Ouen

承包方：Léon Grosse
效果图： Vincent Lafont
7 000 m^2——1998年竣工

一级方程式工厂－PROST GRAND PRIX
凡尔赛(Versailles)，伊夫林省
建筑设计竞赛
建筑用途：高科技赛车样车研发的工厂实验室和办公室
业主：凡尔赛市
业主成员：AFTRP
助理：Jean-Michel Reynier
承包方：Léon Grosse
效果图：Philippe Harden
7 000 m^2——未实施

1998

QUERTIER BOUFFLERS——FERRARE酒店
枫丹白露(Fontainebleau)，塞纳-马恩省
建筑扩建工程：面向宫殿的扩建部分：住宿，商店，商务，办公，地下停车和公用设施用房
业主：枫丹白露市
业主成员：AFTRP
开发商：SEDAF
项目管理：JB/ES & F . Jestaz，E. Legleye
项目组成：Jean-Michel Reynier，Jacques Lévy-Bencheton
景观设计：Bruel/Delmar
项目策划：Dourdin咨询
效果图： Vincent Lafont
20 000 m^2

建筑学院
蒙彼利埃，埃罗省
设计内容：Languedoc-Roussillon建筑学院扩建
业主：文化部
项目组成：Karine Herman，Jérôme Sigwalt
建筑技术：Beterem
工程技术经济：Philippe Talbot
2 800 m^2

BAUER体育馆
圣旺(Saint-Ouen)，塞纳-圣但尼省
建筑设计竞赛
设计内容："红星"新体育馆——位于对已废弃的工业区进行再开发利用的核心区
业主：圣旺市
项目管理：JB/ES & Pierre Rigaudeau
项目组成：G. Bilcandou，Adélaïde Borniche，K. Jensen，Jean-Michel Reynier，N. Soumagnac
承包方：Léon Grosse
效果图： Vincent Lafont
模型：Rod Marawi
25 000 m^2

大学医疗中心
图尔，安德尔和卢瓦尔省
竞赛
设计内容：Trousseau医院扩建
业主：图尔CHU
项目管理：JB/ES & Roger Ivas，Jean Christophe Ballet
项目组成：Vincent Marchand，Gérold Zimmerli
建筑技术：Serete 建设
效果图： Vincent Lafont
模型：L. Salomé
14 000 m^2

国家音乐舞蹈戏剧学院
勒阿弗尔，滨海塞纳省
竞赛——第一名
建筑用途：礼堂，教室，电化教室
业主：勒阿弗尔市
工程主持：Jean-Michel Reynier(竞赛)，Jean-François Bourdet(策划)，Isabelle Vasseur(场地设计)
结构工程师：Terrell 国际——流体设计：Bethac
特效：Changement à Vue
声学设计：Delphi

▲ 大学医疗中心，图尔

UNIVERSITY HOSPITAL CENTRE，Tours

▲ 大学医疗中心，卡昂

UNIVERSITY HOSPITAL CENTRE，Caen

▲ 大学医疗中心，图尔

UNIVERSITY HOSPITAL CENTRE，Tours

▲ BOURGET医院，勒布尔歇

BOURGET CLINIC，Le Bourget

▲ PAUL VALERY大学，蒙彼利埃

PAUL VALERY UNIVERSITY，Montpellier

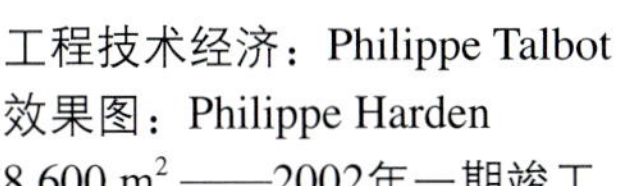

工程技术经济：Philippe Talbot
效果图：Philippe Harden
8 600 m^2 ——2002年一期竣工

勒布尔歇医院
勒布尔歇(Le Bourget)，塞纳-圣但尼省
邀标
设计内容：300床位的医院
业主：卫生总局
项目管理：JB/ES & Gérold Zimmerli
项目组成：Malcolm Nouvel，Jean-Michel Reynier，Vincent Megrot，Jacques Lévy-Bencheton，Philippe Harden
效果图：Didier Ghislain
25 000 m^2

PAUL VALERY大学
蒙彼利埃，埃罗省
竞赛——第一名
建筑用途：戏剧教育建筑，包括剧场，报告厅和教室
业主：教育部，蒙彼利埃大学区区长
执行管理：埃罗省DDE
项目组成：Jean-Michel Reynier，Vincent Megrot，Christophe Kuntz(策划)，Nicolas Crégut(场地设计)
建筑技术：Beterem
工程技术经济：Philippe Talbot
效果图： Vincent Lafont
5 200 m^2 ——2002年竣工

CONTE工厂
滨海布洛涅(Boulogne-sur-Mer)，加来海峡省(Pas-de-Calais)
邀标——第一名
建筑用途：生产制造和贮藏
业主：CONTE
项目管理：JB/ES & J. Jestaz
效果图：Philippe Harden
25 000 m^2 ——未实施

大学医疗中心
图尔，安德尔和卢瓦尔省
竞赛——第一名
设计内容：Clocheville医院改建，建筑空间重新组织——儿科，技术层，病房和药房扩建
业主：图尔CHU
项目管理：JB/ES & Roger Ivas, Jean-Christophe Ballet
工程主持：Vincent Marchand
助理：Jean-Michel Reynier(竞赛)
策划：Tahar Cheref
建筑技术：Sogelerg
效果图：Vincent Lafont
模型：L. Salomé
10 000 m^2 ——2003年竣工

大学医疗中心
卡昂，卡尔瓦多斯省
改建——Clemenceau医院药剂中心扩建
业主：大学医疗中心
助理：Alberto Medem
建筑技术：Serete 建设
3 500 m^2 ——1998年竣工

多媒体文献馆资源中心
Ilot Ségur-Fontenoy，法国就业及社会团结部，巴黎第7区
竞赛——第一名
建筑用途：图书馆，报告厅和问询大厅
业主：法国就业及社会团结部
项目组成：Jean-Michel Reynier(竞赛)，Christophe Kuntz(策划)，Camille Nourrit(家具)，Philippe Vasseur(场地设计)
建筑技术：Sfica
工程技术经济：AEI
3 000 m^2 ——2001年竣工

巴黎第九大学——多芬纳大学
Place du Marechal de Lattre de Tassigny. 巴黎第16区
竞赛——第一名
设计内容：主报告大厅的重新设计
业主：教育部，巴黎第九大学

▲ 巴黎第九大学，多芬纳大学

UNIVERSITY OF PARIS IX–DAUPHINE，Paris

▲ MARCEL DEPREZ高中，巴黎

MARCEL DEPREZ HIGH SCHOOL，Paris

▲ LYON-SUD医疗亭，里昂

LYON-SUD MEDICAL PAVILION，Lyon

▲ 医疗中心，朗布依埃

HOSPITAL CENTRE，Rambouillet

业主代表：G3A
建筑技术：Technip
助理：Philippe Vasseur
1 500 m^2 ——2001年竣工

1999年

MARCEL DEPREZ高中
Rue de la Roquette, 巴黎第11区
竞赛
业主：法国地方政府
业主代表：G3A
项目组成：Jean-Michel Reynier，Vincent Megrot，Jacques Lévy-Bencheton，Lionel Renouf, Camille Nourrit
建筑技术：Sogelerg
效果图： Vincent Lafont
7 000 m^2

劳工部
Ségur-Fontenoy岛，巴黎第7区
竞赛
设计内容：Fontenoy-Ségur岛流通部分重新设计
业主：法国就业及社会团结部
项目组成：Jean-Michel Reynier， Camille Nourrit
建筑技术：Sfica
工程技术经济：AEI
效果图： Vincent Lafont
3 000 m^2

LYON-SUD医疗中心
里昂，罗讷省
竞赛
设计内容：三个医疗中心：心脏病学，内分泌学和肺病学
业主：Civils de Lyon收容救济所
项目管理：JB/ES & Jean-Paul Rouillat
工程主持：Jean-Michel Reynier
助理：Stephanie Boisson
结构工程师：EEG——流体设计：SIRR
工程技术经济：Voutay
效果图：Christophe Valtin-Umlaut
26 000 m^2

医疗中心
德勒(Dreux)，厄尔-卢瓦省(Eure-et-Loire)
竞赛
设计内容：医疗服务设施
业主：Dreux 医疗中心
项目组成：Karine Herman, Jérôme Sigwalt
建筑技术：Setae
6 000 m^2

停车库
索恩河畔沙隆，索恩-卢瓦尔省
邀标——第一名
建筑用途：480各停车位的地下停车场
业主：Parcofrance
助理：Philippe Vasseur
工程主持实施：Lionel Renouf
效果图：Philippe Harden
12 000 m^2 ——2001年竣工

综合体
索恩河畔沙隆，索恩-卢瓦尔省
开发商建筑咨询
设计内容：9个剧院（规模分别为400座、250座和150座），商店，餐馆，住宿，停车
开发商：Dumez-Rhône Alpes/GTM/ADIRA(电影院)
助理：Jacques Lévy-Bencheton
顾问：Claude Fusée
效果图：Philippe Harden
15 000 m^2

医疗中心
德勒，厄尔-卢瓦省
竞赛
设计内容：医疗服务设施
业主：Dreux 医疗中心
项目组成：Karine Herman, Jérôme Sigwalt
建筑技术：Setae
6 000 m^2

医疗中心
朗布依埃(Rambouillet)，伊夫林省
竞赛
设计内容：改建技术层和M.C.O病房

▲ 停车库，索恩河畔沙隆

TOWN HALL PARKING，Chalon-sur-Saône

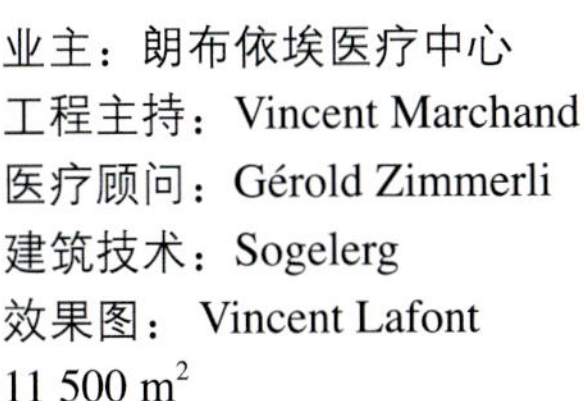

业主：朗布依埃医疗中心
工程主持：Vincent Marchand
医疗顾问：Gérold Zimmerli
建筑技术：Sogelerg
效果图：Vincent Lafont
11 500 m^2

JUSSIEU大学－ESCLANGON建筑
Jussieu大街，巴黎第6区
建筑设计竞赛
建筑用途：为临时教学和研究设立的工业化建筑
业主：EPA Jussieu
业主代表：G3A
承包方：Léon Grosse
工程主持：Jean-Michel Reynier
项目组成：Vincent Mégrot，Jacques Lévy-Bencheton，Lionel Renouf
效果图：Christophe Valtin-Umlaut
9 000 m^2

环保节能中心(ADEME)
昂热，曼恩-卢瓦尔省(Maine-et-Loire)
竞赛——第一名
建筑用途：社会场所，检测楼，高环境标准实施方法和可持续发展原型。办公室，会议室，餐馆
业主：能源环保署

▲ 综合体，索恩河畔沙隆

MULTIPLEX，Chalon-sur-Saône

▲ 环保节能中心(ADEME)，昂热

ENVIRONMENTAL AND ENERGIE SAVINGS AGENCY (ADEME)，Angers

业主代表：SARA
项目策划：CPO策划——人类工程学——Tribu
工程管理：JB/ES & Jean-Pierre Bastide
工程主持：Jean-Michel Reynier
项目组成：Isabelle Vasseur，Vincent Mégrot，Marie-Hélène Torcq
景观设计：Bruel Delmar
环境事务合作：Pierre Diaz-Pedregal
建筑技术：Séchaud，Bossuyt
效果图：Vincent Lafont
10 000 m^2——2002年竣工

办公楼
圣克洛大街，凡尔赛，伊夫林省
业主：Leon Grosse
助理：Isabelle Vasseur
效果图：Vincent Lafont
4 000 m^2——2002年竣工

MAISON BLANCHE精神病医院
Rue d' Haureville，巴黎第10区
竞赛
设计内容：Saint Martin部分(60床)，La Tour d' Auvergne(40床)
业主：Maison Blanche公共建设委员会
业主代表：SCIC发展
工程主持：Jean-Michel Reynier

▲ 办公楼，圣克洛大街，凡尔赛

OFFICES，Avenue de Saint-Cloud, Versailles

▲ MAISON BLANCHE精神病医院，巴黎

MAISON BLANCHE PSYCHIATRIC HOSPITAL，Paris

助理：Vincent Mégrot
建筑技术：Setae
工程技术经济：Philippe Talbot
6 000 m^2

国家农业结构组织中心(CNASEA)
利摩日，上维埃纳省
竞赛
设计内容：利摩日CNASEA中心的内部重新布置：办公室，计算机中心
业主：农业部/CNASEA
业主代表：G3A
工程主持：Jean-François Bourdet
项目组成：Astrid Rappel，Alexis Lorch，K. Berger
建筑技术：Séchaud，Bossuyt
效果图：Eric Anton
模型：Alpha Volumes
10 000 m^2

办公楼
圣克莱尔市，卡尔瓦多斯省
空间设计：Shema
业主：SOGEA
工程主持：Jean-Michel Reynier
助理：Jean-François Bourdet
3 000 m^2

▲ 国家农业结构组织中心，利摩日

NATIONAL CENTRE FOR AGRICULTURAL STRUCTURES，Limoges

▲ Jussieu大学，巴黎

JUSSIEU UNIVERSITY，Paris

▲ NATEXIS宴会中心，沙伦顿

NATEXIS BANQUES POPULAIRES，Charenton

▲ HAUT-ANJON医疗中心，贡捷堡

HOSPITAL CENTRE OF HAUT-ANJON，Château-Gontier

▲ ILOT VILLOT-RAPEE住宅，巴黎

ILOT VILLOT-RAPEE HOUSING，Paris

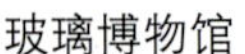

玻璃博物馆
萨尔波特里(Sars-Poteries)，北部省
建筑用途：工作室，展览空间
工程主持：Jean-Michel Reynier
效果图：Philippe Harden
800 m^2

2000年

NATEXIS宴会中心
沙朗通勒蓬(Charenton-le-Pont)，瓦勒德马恩省
邀标
设计内容：管理部门，会议中心，餐馆
业主：Natexis宴会中心
工程主持：Jean-Michel Reynier，Vincent Marchand
项目组成：Jacques Lévy-Bencheton，Jean-François Bourdet
建筑技术：Philippe Talbot/Bethac
安全顾问：Casso/Seet Secoba
效果图：Eric Anton/Vincent Lafont/Laurent Hocherg
模型：Alopha Volumes / Rod Marawi
5 300 m^2

欧莱雅会议中心
Aulnay-Chanteloup，塞纳-圣但尼省
设计内容：会议中心，餐馆——俱乐部，场地重新规划
业主：欧莱雅
流体力学：Bethac
工程技术经济：Philippe Talbot
效果图：Laurent Hochberg
模型：Alpha Volumes
5 700 m^2

"LES MASSUES"医疗－外科手术中心
里昂，罗讷省
邀标
设计内容：医疗－外科手术中心改扩建
业主：Groupama 保险
项目管理：JB/ES & Jean-Paul Rouillat
建筑技术：Sodeteg
效果图：Martignac
14 000 m^2

织物中心
里昂，罗讷省
竞赛
设计内容：罗纳阿尔卑斯地区织物中心(突出它创造性、发展性、交流性)
业主：UNITEX
业主代表：SERC
项目管理：JB/ES & Jean-Paul Rouillat
助理：Jean-Michel Reynier
建筑技术：Matté s. a.
工程技术经济：Cabinet Denizou
效果图：Laurent Hochberg
3 800 m^2

ILOT VILLOT-RAPEE住宅
Quai de la Rapee，巴黎第12区
竞赛
设计内容：70个住宅单元
业主：巴黎OPAC
助理：Jean-Michel Reynier
效果图：Didier Ghislain
5 000 m^2

HAUT-ANJON医疗中心
贡捷堡(Château-Gontier)，马耶讷省(Mayenne)
竞赛
设计内容：技术层和150床医院
业主：医疗中心
业主代表：DDE
项目管理：JB/ES & Willam Gohier
项目组成：Vincent Marchand，Vincent Megrot，Christophe Kuntz，Jacques Lévy-Bencheton
医院建筑顾问：Gerold Zimmerli
景观设计：Alfred Peter
结构和技术经济工程师：AUA结构——
流体设计：SETAE
效果图：Didier Ghislain
模型：Alpha Volumes
20 000 m^2

▲ 医疗中心，戛纳

HOSPITAL CENTRE，Cannes

▲ FACULTE D'ASSAS，巴黎

FACULTE D'ASSAS，Paris

▲ WARSAW 法国驻波兰大使馆，贡捷堡

FRENCH EMBASSY，Warsaw (Poland)

▲ 医疗中心，圣日耳曼昂莱

HOSPITAL CENTRE，Saint-Germain-en-Laye

▲ AWON办公楼，马西

AWON OFFICES，Massy

医疗中心
利摩日，上维埃纳省
竞赛
妇幼中心
业主：利摩日医疗中心
项目主持：Vincent Marchand，Jean-Michel Reynier，Tahar Cheref
建筑技术：Terrell国际——流体设计：Bethac
工程技术经济：Philippe Talbot
效果图：Vincent Lafont / Laurent Hochberg
20 500 m^2

医疗中心
圣日耳曼昂莱，伊夫林省
竞赛
设计内容：妇幼中心，改建
业主：Poissy社区医疗中心/圣日耳曼昂莱
业主代表：SCIC发展
项目组成：Hélène Planex，Vincent Mégrot
建筑技术：SETAE
效果图：Philippe Harden
9 200 m^2

医疗中心
戛纳，滨海阿尔卑斯省
竞赛——第一名
业主：Cannes医疗中心
业主代表：SCIC发展
项目管理：JB/ES & Associes建筑师工作室(Gilles Gauvain)
医院顾问：Gérold Zimmerli
建筑技术：Terrell Rooke Associés－设备供应商：SIRR
工程技术经济：Philippe Talbot
效果图：Vincent Lafont
模型：Alpha Volumes
55 000 m^2——2006年竣工

FACULTE D'ASSAS
Rue d'Assas，巴黎第6区
竞赛
业主：教育部
业主代表：SCAP
结构工程师：Terrell Rooke Associes—流体设计：Béthode
工程技术经济：G. V. Ingénierie
OPC：Gémo
效果图：Vincent Lafont

生物中心
里尔(Lille)，北部省
竞赛
建筑用途：实验室和办公室
业主：CHU de Lille
项目组成：Jean-François Bourdet，Vincent Mégrot
建筑技术：SETAE
工程技术经济：Philippe Duval
效果图：Vincent Lafont
模型：Alpha Volumes
18 000 m^2

法国大使馆
华沙，波兰
竞赛——评委会奖
业主：外交部
项目组成：Jean-Michel Reynier，Astrid Rappel, Jacques Lévy-Bencheton, Lionel Renouf
建筑技术：Séchaud & Bossuyt / Bief
项目策划：Dourdin 咨询
效果图：Eric Anton
模型：Alpha Volumes
6 000 m^2

AWON办公楼
马西(Massy)，埃松省(Essonne)
邀标——第一名
业主：Awon团队
项目组/业主：M.A.P.S
结构工程师：Terrell Rooke Associés——流体设计：Béthode
工程技术经济：Philippe Talbot
12 000 m^2

附 记
Appendix

助理

Marie AKAR, Marine ALEXANDRE, Frederic ALLI-GORIDES, Christele ANDRIEUX, Eduardo ARROYO, Marianne ARTARIT, Eric BABIN, Catherine BAI-BRUNET, Laurent BANDINI, Christian BASSET, Christophe BAUBIL, Matthias BAUER, Elisabeth BEAULIEU, Martin Gonzalo BENAVIDES, K. BER-GER, Veronique BEYRON, Herve BLETON, Gilles BOCION, Stephanie BOISSON, Adelaide BOR-NICHE, Jean-Frangois BOURDET, Laurent BOU-THONNEAU, Gilles BRESARD, Colette BRICE, Nathalie BRILMAN, Claire BRUNET, Veronique BRUNET, Aline CABOURET, Sandrine CARDON, Marc CHASSIN, Guillemette CHENIEUX, Tahar CHEREF, Nicolas CHEVRON, Jean-Christophe CHOTTIN, Thomas CORBASSON, Elsa CORTESSE-VINCENT, Eric CUGNART, Myriam DAO, Sophie de RENZY-MARTIN, Alain DEBORD, Emma-nuelle DELABY, Gabriel DELAGE, Igor DEMIDOFF, Brune DESALLAIS, Sylvestre DIDIER, Wolfgang DIEVERNICH, Ingrid DOM, Cecile DOMAIN, Catherine DORMOY, Pierre Emmanuel DROSTE, Eric DUFOUR, Herald DUPLESSY, Marie-France DUS-SAUSSOIS, Mounia ELGHARBOUJ, Sandra ELIAS-ZWICZ, Serge ELOIRE, Catherine EMONET, Pascal EUGENE, Vincent FALCUCCI, Valerie FAUGERAS, Dominique FELDSTEIN, Patricia FERARU, Nicolas FONTY, Daniele FORTE, FRIGO, Victor FUENTES, Fred GADAN, Francis GALLOIS-MONTBRUN, Christine GHIAI-FAR, Frangois GILLET, Piotr GLE-GOLA, Edouard GRASSIN, Christophe GREA, Alain GRIGNET, Daniel GUETTA, Denis GUFFROY, Nicolas GUICHET, Francois GUIDON, Frederique GUIL-LOUET, Damien HAMON, Brian HEMSWORTH, Karine HERMAN, Noe HERNANDEZ, Christiane HERVET, Thomas HEUZE, Anne-Marie HORS, Enrico INGALA, Youssef ISMAEL, Ariane JOUAN-NAIS, Laurent JOUANNEL, Claire KARSENTY, Laurent KARST, Pawel KORNECKI, Christophe KUNTZ, Christine LAEMLIN, Xavier LAGURGUE, Thierry LANDI, Herve LANGLAIS, Clare LASBREY, Philippe LE, Jacques LE DU, Sophie LE PICARD, Elisabeth LEMERCIER, Herve LEVASEUX, Pierre LEVAVASSEUR, Jacques LEVY-BENCHETON, David LONG, Alexis LORCH, Lionel LORIS, Babak MAD-DAH, Vincent MARCHAND, Bastien MARION, Renaud MARRET, Federico MASOTTO, Alberto MEDEM, Vincent MEGROT, Stanislas MELUN, Nicolas MEMI, Sophie MERTIAN, Christophe MICHEL, Maria-Grabriela MOLISE, Bach NGUYEN, Sophie NICOLAS, Mojan NOUBAN, Camille NOURRIT, Malcolm NOUVEL, Cathy OLIVE, Jean-Marie PARANT, Eric PARISIS, Eddy PENOT, Ra-phaels PERRON, Veronique PETIT, Vitorio PISU, Sabine PLANCHOT, Helene PLANEX, Michel PLA-XINE, Gerard PRAS, Anna PRICOUPENKO, Sophie PRUDHOMME, Astrid RAPPEL, Maya RAVEREAU, Jean-Jacques RAYNAUD, Nathalie REGNIER, Jean-Francois RENAUD, Lionel RENOUF, Jean-Michel REYNIER, Line SATTLER, Nathalie SCALI-RING-WALD, Toshio SEKIGUCHI, Arnaud SENESI, Genevieve SHATER, Hayan SIDAOUI, Jerome SIGWALT, N. SOUMAGNAC, Samy TALAOUBRID, Annemick THOMASSEN, Marie-Helene TORCQ, Jean-Fran-gois TROULARD, Simone TROUVE, Is.abelle VAS-SEUR, Philippe VASSEUR, Helene VERNET, Matthieu VIEDERMAN, Elsa VINCENT-CORTESSE, Thomas VON AMELUNXEN, Claudia WENDER, George YIONTIS, Joanne YOUNG, Sylvaine ZABO-ROWSKI

照片

Herve ABADIE, Nicolas BOREL, Georges FESSY, Jean-Marie MONTHIERS, Thierry O' SUGHRUE

模型

ALPHA-VOLUMES, Eric CHICAEZ, Jean-Louis COURTOIS, Eric DELARUE, Olivier DOIZY, Jean-Michel FRANCOISE, Bruno ISAMBERT, Rod MARANI, David TOPANI

艺术

Alberto BALI, Virginie CAMPIQN, Louis DERBRE, GATIMALAU, Vincent GONTHIER, Frangois MORELLET, Anna PRICOUPENKO, Francois SEIGNEUR, Anton SOLOMOUKA

景观

David BESSON-GIRARD, Anne-Sylvie BRUEL, Christophe DELMAR, Kathryn GUSTAFSON, Karine HERMAN, Florence MERCIER, Philippe NIEZ, Michel PENA, Alfred PETER ROZOT, Alexandra SCHMIDT, Gilles VEXLARD

顾问

Jean-Pierre AURY, Christian BARTENBACH, Georges BERNE, Jean-Louis BESNARD, Sophie BRINDEL-BETH, Daniel DAR-BOIS, J.P. DESCHAMPS, Pierre DIAZ-PEDREGAL, Jerome DOURDIN Claude FUSEE, Philippe LAIR, Marc MALINOWSKI, Pierre MIQUEL, Alain PAPI-NEAU, Richard PEDUZI, Sabine ROSANT, Jean-Paul ROYNETTE, Guillaume SAALBURG, Philippe UZZAN,GeroldZIMMERLI

透视图

Jacques AILLAUD, Eric ANTON, Antoine BONO-MO, Anne et Denis CLEARY, Didier GHISLAIN, Philippe HARDEN, Douglas HARDING, Laurent HOCHBERG, Thierry LACOSTE, Vincent LAFONT, Louis PAILLARD, Francois ROBAIN, Jerome SIGWALT, Frederic TERREAUX, Christophe VALTIN

合作实践

AR&C - Pierre Rigaudeau et Philippe Coeur, AVANT-TRAVAUX, Jean-Pierre BASTIDE, DELAGE et TSAROPOULOS, Gilles GAUVAIN, William GO-HI ER, IVARS et BALLET, Francois JESTAZ, LAM-POTANG et SIEW, Eric LEGLEYE, Dominique PERRAULT, Jean-Paul ROUILLAT, Jean-Michel WIL-MOTTE

图片提供

Jean-Marie Monthiers Cover, 20, 30, 48, 52, 53, 54, 62, 75, 76, 77, 78, 93, 94, 95, 96, 97, 98, 105, 106, 107, 108, 109, no, 113, 115, n6

Herve Abadie 6, 14, 36, 60, 65, 67, 68, 119, 120, 121, 122

Dimension color 26

Nicolas Borel 18, 87, 88, 89, 90

Vincent Lafont 32

F. Achdou/Urba Images 34/35

Brunet/Saunier 38, 40, 42, 44, 46, 114

Georges Fessy 58, 72, 73, 8i, 82, 83, 84, 101, 102

IGN 64, 70, 74, 80, 86, 92, 100, 104, 112, 118

Thierry O'Sughrue 66